THE ROCKS DON'T LIE

Also by David R. Montgomery

Dirt: The Erosion of Civilizations

King of Fish: The Thousand-Year Run of Salmon

THE ROCKS DON'T LIE

~~~~

A Geologist Investigates Noah's Flood

~~~~

David R. Montgomery

W. W. Norton & Company
New York London

Printed in the United States of America
First Edition

For information about permission to reproduce selections from this book,
write to Permissions, W. W. Norton & Company, Inc.,
500 Fifth Avenue, New York, NY 10110

For information about special discounts for bulk purchases, please contact
W. W. Norton Special Sales at specialsales@wwnorton.com or 800-233-4830

Manufacturing by RR Donnelley, Harrisonburg, VA
Book design by Kathryn Parise
Production manager: Anna Oler

LIBRARY OF CONGRESS CATALOGING-IN-PUBLICATION DATA

Montgomery, David R., 1961–
The rocks don't lie : a geologist investigates Noah's flood /
David R. Montgomery. — 1st ed.
p. cm.
Includes bibliographical references and index.
ISBN 978-0-393-08239-5 (hardcover)
1. Paleolimnology. 2. Paleohydrology. I. Title.
QE39.5.P3M66 2012
551.48'9—dc23
2012015146

W. W. Norton & Company, Inc.
500 Fifth Avenue, New York, N.Y. 10110
www.wwnorton.com

W. W. Norton & Company Ltd.
Castle House, 75/76 Wells Street, London W1T 3QT

2 3 4 5 6 7 8 9 0

For my parents, Dave and Toby,

with thanks for encouraging me to think.

Contents

Preface

~~~~

All around the world, mythology and folktales address the origin of topography, the form of the land itself. How should we read ancient stories, like accounts of great floods purporting to explain the origin of landforms? Can we regard them as tales of prehistoric events, or should we dismiss them as archaic superstition? As a geologist trained to read the history of the world from rocks and landforms, I'm curious about the geological basis of folktales and how geography, culture, and tradition shape the way people see and interpret the land.

Investigating the origin of the world's flood stories, some of humanity's oldest and most widely spread traditions, presents an intriguing challenge. Geologists tend to explain the prevalence of flood stories among ancient societies as simply reflecting the fact that floods are common natural disasters. But could there be more to stories of really big floods, or even the flood to end all floods, Noah's Flood?

Of all the sciences, geology is especially bound to the story of
~~~~

Noah's Flood. Historically, few things on the frontier between science and religion proved as contentious as the biblical stories of the Creation and Noah's Flood, the age of the world, and the genesis of topography. For two centuries, Christians have wrestled with contradictions between traditional biblical interpretations and geological discoveries. At the same time, debate over interpreting supposed signs of Noah's Flood made surprising contributions to the development of geology. This back-and-forth also proved central to the rise of modern creationism and its perception of geology as a fundamental threat to faith.

I started writing this book intending to present a straightforward refutation of creationism, the belief that the world is a few thousand years old and that all the world's topography—every mountain, hill, and valley—was formed by the biblical Flood. But as I read through old books I learned how stories about enormous floods shaped both scientific and religious views. I also came upon a different story about the nature of faith.

In looking into the origins of flood stories, and the story of Noah's Flood in particular, I thought I'd find the standard conflict between reason and faith. Instead, I found a much richer story of people struggling to explain the world—and our place in it. The initial development of the discipline of geology was premised on the Flood as fact, which naturally led to imaginative theories of how to interpret the story of Noah's Flood. Later, with evidence literally in their hands and beneath their feet, geologists began to influence theology, showing that a global flood fell short when tested against the rocks that make up our world. Along the way, scientists were as apt to be blinded by faith in conventional wisdom as Christians proved adept at reinterpreting biblical stories to account for scientific findings. The historical relationship between science and religion was far more fluid, far more cross-pollinating than I ever thought—or was taught at Sunday school or in college.

Little did I expect to learn that Niels Stensen, also known as Steno, the seventeenth-century grandfather of modern geology, invoked Noah's Flood to explain the origin of the landscape around Florence in his influential treatise on the nature of fossils. All too frequently, the history of science is simplified into a story of the light of reason dispelling the shadows of myth and superstition.

I was equally surprised to learn that the development of modern creationism originates in arguments within the fundamentalist community over how Noah's Flood explained geology. I did not expect to learn that the historic interplay between how Christians interpreted biblical stories and how scientists continually reinterpreted geological evidence helps explain the origin of modern creationism and why such beliefs arose in America. Neither did I realize that what we know today as creationism is one of the most recently evolved branches of Christianity, or that the founding fathers of modern creationism based their views, in part, on a perceptive critique of geology in the days right before the discovery of plate tectonics. Modern creationists had a rational basis for their arguments—even if they then recycled thoroughly discredited seventeenth-century theories to support their beliefs.

For readers seeking to delve deeper into the history of science and religion, and the topics touched upon in this book in particular, I have listed my sources at the end. I leave it to others to debate the long and fruitless search for Noah's ark and where it came to rest (Mount Ararat being a relatively recent addition to a long list of candidate sites). I will also steer clear of arguments over the size, design, and logistics of the ark, despite all the wonderfully imaginative ideas about how to accommodate a world of animals on a handmade lifeboat. And I leave debate about the question of intelligent design to theologians in whose purview inherently untestable ideas properly reside. While my geological background and training provide me with insight into earth history—how to read stories archived in stone and etched upon the land—I have no better idea than anyone why the universe exists and runs the way it does. Such questions are unanswerable, at least in this lifetime.

I found reading the historic works of theologians, natural philosophers, and scientists a fascinating experience, one that left me with an appreciation for the rich and engaging interplay between biblical interpretation and the development of geology. Noah's story is central to one of the longest-running debates between science and religion as people sought, and still seek, to reconcile scriptural interpretation with observations of the natural world. We mortals have long been struggling to understand who we are, and probably always will. Even today, interpretations of the biblical flood story remain central to understanding modern culture wars—no matter how one views them—because how we read ancient stories still defines the way we see the world, and thus ourselves.

THE ROCKS DON'T LIE

1

Buddha's Dam

As a geologist, I've had plenty of surprises in the field, but I never expected that an excursion to a remote corner of Tibet would lead me to a new appreciation for the biblical story of Noah's Flood. My specialty is geomorphology, the study of processes that create and shape topography. Over the last several decades I've explored how landscapes evolve—where stream channels begin, how landslides sculpt hillslopes, and why rivers carve deep gorges through mountain ranges.

In the spring of 2002 I joined a research expedition to the Tsangpo River in southeastern Tibet. The team needed a geomorphologist with river experience to study how the Tsangpo had sawed down through kilometers of rock to carve the world's deepest gorge. I couldn't turn down the chance to visit the roof of the world.

As we drove down from the pass toward the Tsangpo on the newly paved road southeast of Lhasa, I noticed flat-topped piles of sediment rising above the valley floor. Known as topographic terraces, these elevated islands of flat ground can form in different ways,

Photograph of a topographic terrace, the top of which corresponds to the level of an ancient lake that once filled the valley of the Tsangpo River, Tibet (photograph by the author).

most commonly when an incising river abandons an old riverbed. I watched for clues to determine what created these.

Over the course of our several-week expedition I collected the pieces of a landscape-scale puzzle. Flat-topped piles of loose sediment—gravel, sand, and silt—stuck up hundreds of feet into the air where tributary valleys entered the main valley. More terraces lay at about the same elevation at each confluence where smaller streams joined the river. From our hotel near the foot of the valley wall we could see a terrace rising above the edge of town a few blocks away. A short hike up a dirt road cut up through the side of the terrace revealed hundreds of alternating layers of silt and finer clay. Segregation into distinct layers sorted by size meant that these sediments were laid down in quiet water. Such fine material would not have settled out in a turbulent river. The implication was clear. An ancient lake once filled the valley.

Sketching the extent of the flat terrace surfaces onto our map as we drove along the valley, I badgered my compatriots into occasionally stopping for a closer look at these curious piles of sediment laid out like a giant's playground. Some were dried-up gravel riverbeds perched well above the modern river. Others were lake terraces made of layered silt and clay. How did they get there?

A coherent picture began to emerge as we traversed up and down the valley. The terraces made of river gravel continued down the valley bottom to an elevation corresponding to the top of the lake sediments, defining the ancient shoreline where the rivers had entered the lake. In addition to prominent terraces that rose a few hundred feet above the modern river, the remnants of a second set of higher terraces preserved at a few remote locations halfway up the valley walls attested to an even deeper lake. At least twice in the recent geologic past a lake extended hundreds of kilometers upstream from the Tsangpo Gorge. I was onto something.

It was thrilling to have scientific sleuthing that started as little more than a hunch lead to a solid story. Once I saw the pieces and knew how they fit together, the story of ancient lakes that once filled the valley of the Tsangpo stood out plain as day in the form of the land.

What do you see when you look at the land? Something stable and reassuringly solid. A slope to ski down? A surface to pave over? Geologists see the world as incredibly dynamic and ever-changing—only change occurs slowly over immense spans of time.

I've learned to see what the land used to look like, and what it might look like in the future. Reading a landscape is an ongoing process of combining curiosity and inquiry. Why is that hillside bare and rocky? Why is that one covered with soil? Deciphering topography makes geologists natural storytellers. We piece together fragmentary clues in rocks and landforms to connect dots across landscapes, mountain ranges, and continents and tell stories with whole chapters

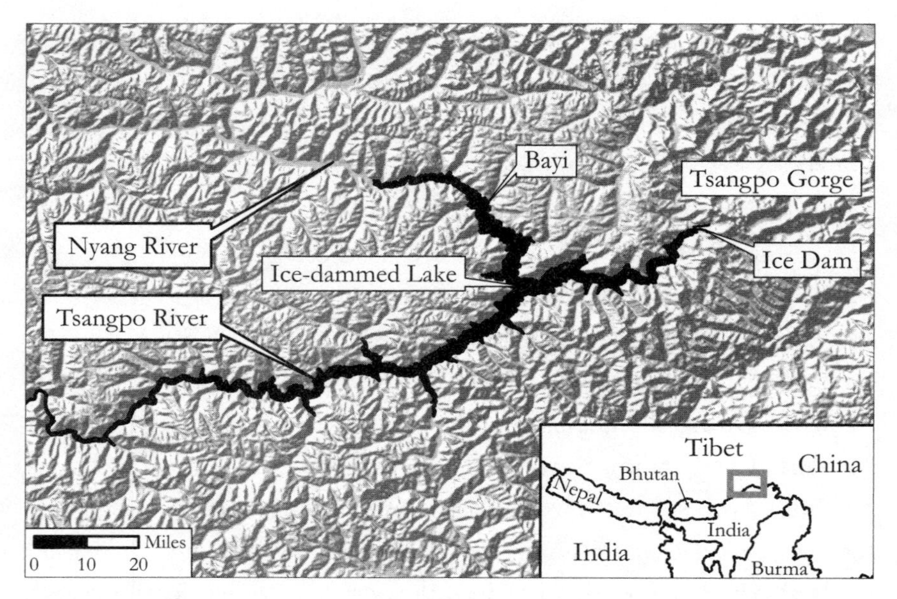

Map of the Tsangpo River, Tibet, showing the Nyang River, the town of Bayi (where our hotel was located), and the moraine dam at head of Tsangpo Gorge. The reconstructed extent of the lower paleo-lake is shown in black.

lost to erosion and time. And here in the valley of the Tsangpo was a great story, except for one big loose end.

Looking at the map, there was no obvious dam to hold back our newly discovered ancient lakes. What kept them from draining down into the Tsangpo Gorge? Many miles downstream, right at the head of the gorge, we discovered glacial debris plastered on both sides of the valley confirming that a massive tongue of ice had once plunged down the 25,000-foot-high peak of Namche Barwa and blocked the river. The two levels of terraces extending far upstream indicated that a wall of ice and mud dammed the river, not just once but time and again, backing up a great lake that filled the valley.

As you might imagine, ice doesn't make a very good dam. Once the lake filled enough to float or breach the dam, a rush of liberated water roared down the gorge, scouring out everything in its path. Upstream of the gorge, we found horizontal stripes of silt plastered onto the valley walls. Here were old shorelines confirming that the ice advanced to block the river over and over again, most likely during cold glacial periods or at times when strong monsoons delivered extra snow to the glacier's source on the high peak. As the glacier repeatedly dammed the throat of the gorge, ice-dam failures generated catastrophic floods that drained ancient lakes impounded in the valley upstream.

One day as we drove down the valley toward the gorge, one of my graduate students relayed information from a guidebook he'd brought along. Local folklore told of a traditional kora—a Buddhist pilgrimage trek—that circled a small peak ringed by lake terraces. Pilgrims walk the kora to commemorate how Guru Rimpoche brought Buddhism to Tibet through defeating a powerful lake demon, draining its home to reveal fertile valley-bottom farmland. It was a feat impressive enough to convert the locals. I began to think that an oral tradition might record our glacial dam-break flood.

Suspicion moved beyond idle speculation when we got the radiocarbon dates from wood fragments I'd painstakingly collected from the lake terraces. Radiocarbon dating uses the ratio of carbon isotopes in once-living matter to determine how long ago it died.[1] The technique works because carbon-14 (^{14}C) decays to carbon-12 (^{12}C) at a known rate, and all living things start with a $^{14}C/^{12}C$ ratio equal to that in the atmosphere from which the carbon was originally taken up by photosynthesis. Wood fragments from the higher terraces of the older lake were almost ten thousand years old, dating from the tail end of the last glaciation of what's popularly known as the ice age. Fragments from the younger lake were only about twelve hundred years old—dating from around the eighth century AD.

This was about the time that Guru Rimpoche arrived in Tibet. Did the geologic story I read in the landforms really support a Tibetan folktale? Or was it that the folktale told the geologic story?

Two years later, in 2004, I returned to Tibet to explore the story of the lake-draining flood. On our first trip we had hired a local farming couple to collect monthly samples of river water. When I told the farmer's wife how we'd discovered that the whole valley was once an ancient lake, she replied that yes, she knew about that. Caught off guard, I listened to her. She pointed out a steep hillside across the valley and described how three boats had been stranded there as the lake drained to reveal the farmland of the modern river bottom. She told me she'd heard the story from the Lamas at the local temple.

The temple sat right on top of a stack of ancient lake sediments, one of the terraces that rose to the elevation of the lower lake shoreline. A painted temple wall along an exterior walkway even had a striking portrayal of Guru Rimpoche above a lake floating before the distinctive peaks flanking the gorge entrance. Asked how Guru Rimpoche

Photograph of the truncated glacial moraine where an ancient ice dam extended down off the flanks of Namche Barwa (the high peak in the background at right) to dam the Tsangpo River immediately upstream of the Tsangpo Gorge (photograph courtesy of Bernard Hallet).

had drained the lake, the head Lama said he cared not *how* the great master did it. What was important was the fact that Guru Rimpoche had. Besides, he continued, the story we should be interested in was how the ocean once covered all of Tibet. He described how he had seen water-rounded rocks perched on mountainsides high above the valley. He assured us that the ocean once covered the high peaks. His story of a flood that submerged the world sounded familiar.

People around the world tell stories to explain distinctive landforms and geological phenomena. The global distribution of folklore associated with topography and great floods makes me suspect that people are hard-wired to be fascinated by and to question the origin of landscapes. I know I am. I think I was a geomorphologist before I knew what one was.

As a kid I could stare at maps for hours, examining geographical details of places I'd never been. Seduced by the lay of the land, I grew up intrigued by how topography shaped historic battles, controlled the locations and form of cities, and forged the character of civilizations.

As a Boy Scout I loved hiking in the Sierra Nevada and the coastal mountains of California. I would follow our treks matching map to landscape, tracking the progress we'd made. Because I rarely got lost, my parents designated me navigator on family vacations through the wide-open landscapes of the American West. Driving through country with rocks laid bare or sculpted into cliffs, spires, and mesas held my attention. I would chart our route on a map, carefully noting where we were, my head swinging from the map on my lap to the landmarks out the car window and back again. What river were we following? What range were those mountains on the horizon? My love of topography and maps—my topophilia—cultivated an eye for understanding landscapes.

It wasn't until college that I learned to recognize and name the processes behind why Earth's surface looks as it does, and how to read the signature of erosion and deposition in shaping landscapes. Most people see the land as static. I learned to see it as ever-changing.

Now, wherever I am in the world I look to the shape and arrangement of hillslopes and valleys, mountains and rivers, to read the processes that shaped the land. There is something inherently beautiful about topography, in the rhythmic rise and fall of rolling hills, a soaring wall of rock rising to a rugged mountain peak, or the looping symmetry of a great river meandering across a wide-open floodplain. Coming to understand the forces that sculpt our world has nurtured the sense of wonder and beauty I find in nature. I've also found in my travels and expeditions that, like me, people all over the world are enthralled with and tell stories about topography.

Some of humanity's oldest stories are about the origin of the world and its landforms. Why do volcanoes exist? How did the oceans form? When did it all begin? People have wondered about such things for about as long as they've been thinking. How am I sure of this? We live on Earth's surface, and the lay of the land influences almost everything we do, all the more so if you've ever climbed a mountain, or found yourself in a flood, an earthquake, or near an exploding volcano. How the world was made and how it works is of interest to anyone living on Earth—which pretty much covers everybody.

After Tibet, another encounter with flood traditions made me suspect that there may be more truth to flood stories than I ever imagined. A bright spot in the tragic December 2004 tsunami that decimated Indonesia and Thailand was the remarkable tale of how the Moken people, the region's sea gypsies, survived without casualties because they knew to run for the hills. These seafaring people had an oral tradition of a big flood that warned them to get to high ground when the tide mysteriously went out far and fast. Knowledge that it would soon come back in as a monstrous wave helped them survive, and gave them a chance to pass the story to future generations.

Could science be playing catch-up to folklore? For most of our history as a species, oral traditions were the only way to preserve knowledge. So why wouldn't the world's flood stories record actual ancient disasters? After all, the world's first civilizations were agricultural societies settled along major floodplains where swollen rivers periodically submerged fields and towns. And, of course, among the best-known and most controversial flood stories is that of Noah's Flood. Could there also be truth to the biblical tale?

Today, geologists generally dismiss Noah's Flood with a chuckle and shrug it off as a relic of another time. But for centuries it was considered common knowledge among Christians and many natural philosophers that Noah's Flood shaped our world. What else could have? If the planet itself was but a few thousand years old, as Christians believed the Bible implied, how could the processes we see today have possibly shaped a mountain like Everest or a place like the Grand Canyon, let alone the whole world? The work of rivers slowly grinding away at a mountain range would not add up to much even over dozens of centuries. The busted-up rocks and rough terrain of mountains were seen as the ruins of a former, once perfect world that raging floodwaters destroyed at the behest of an angry god. Topography was concrete evidence of the awesome power of divine wrath, a humbling reminder of our place in the grand scheme of things.

Throughout history, stories about catastrophic floods have been framed by conflict between orthodoxy and heresy—both religious and scientific. At first, arguments from all sides assumed that the best theories were those that could predict what was not yet known. Answers to the question of how to read the land lay rooted in how to interpret physical evidence one could bash open, kick over, or dig down into to test ideas about what should be there. Theories could be tested against evidence.

In Sunday school I learned that Bible stories were parables to be read more for their moral message than their literal words. The story of Noah's Flood taught mankind to be stewards of the environment—to care for all parts of nature, even as we bent her to our desires. Growing up, I was satisfied that Jesus taught how to live a good life and that science revealed how the world worked.

Through all my schooling I never thought much about conflict between science and religion. Then, in my early thirties, I met a gregarious fundamentalist on jury duty. While I was waiting to be called for jury selection, a middle-aged woman sitting next to me snuck a peek at the paper I was reading and tried to strike up a conversation: "Isn't it amazing how Mount Saint Helens shows Noah's Flood carved the Grand Canyon?"

I looked up, roused from an account of how the rivers draining the volcano's flanks carved deep canyons into loose debris after the eruption. She continued, asking me if I recalled how many thousand years ago Noah's Flood had reshaped the world. My raised eyebrows and open mouth probably telegraphed my thoughts. When I told her that a global flood was pure fiction and suggested that she might want to tack a few more zeros onto the planet's age, she responded that only atheists believed the world was ancient. I sat there at a loss for words—something geology professors are not generally known to be. A loudspeaker calling her to jury service ended our awkward conversation.

My jury-duty mate is not alone in believing that Noah's Flood explains nearly all of earth history. Her view is what geologists call "flood geology," the resilient yet scientifically discredited idea that the biblical flood remodeled the planet in one fell swoop several thousand years ago. In the four hundred years since the church grounded Galileo, Christianity has grown to accept science that disproves archaic notions about our world being the center of the universe. Why should geological discoveries be treated any differently than those of astronomy?

The more I looked at the history of efforts to explain Noah's Flood, the more I came to realize that our cultural view of a centuries-long, ongoing conflict between geology and Christianity—between science and religion—was too simplistic. The real story was far more interesting.

My curiosity about a geological basis for the biblical flood began in the 1990s, when Bill Ryan and Walter Pitman, two prominent oceanographers, suggested that the Mediterranean Sea catastrophically spilled over into a low-lying lake valley to create the Black Sea. When they proposed that this was in fact Noah's Flood, many Christians were intrigued by scientific support for the biblical story. Creationists were outraged. Why were creationists angry when scientists claimed to find support for the biblical flood? The problem to them was that this flood was not an earth-shattering, topography-busting flood that ripped apart and reassembled the whole world. It was not the flood that they thought the Bible described. They saw the suggestion that Noah's Flood was a regional disaster, and not a global event, as an attack on Christianity. For completely different reasons, many geologists also were immediately skeptical—hadn't science dispelled Noah's Flood as an ancient myth?

I thought Ryan and Pitman's idea made sense. It was geologically plausible. Had they solved the puzzle of Noah's Flood?

No other story has had as profound an influence on geology as that of Noah's Flood. Today almost half of the American public believes in young Earth creationism—that the world is about six thousand years old and that Noah's Flood reshaped Earth's surface into today's world a few thousand years before the time of Christ.[2] While there is no doubt that the world is far older than creationists allow, it is this most fundamental feature—time—eons of it, that causes creationists to so vociferously deny modern geology. Why this reaction? Because if the world is old, it allows time not only for mountains to rise and erode but, more problematically, for evolution to work. In defending an interpretation of God's word contradicted by geological evidence, creationists abandon a long-standing Christian belief that rocks don't lie.

For centuries, Christians interpreted scientific discoveries through faith that God's word (the Bible) and Creation (nature) must be consistent with one another. In combing through historical material—both geological and theological—I saw that previous generations had reconciled geological evidence with how to read the biblical story of Noah's Flood. Although harsh rhetoric is by no means a modern invention, for centuries few considered science and religion mortal enemies. Most early geologists were clergy who believed that stories preserved in rock revealed the nature of God's works as surely as the Bible revealed His Word. Scientifically inclined clergy had faith that discoveries about the natural world would illuminate biblical interpretation. They knew their efforts could only enhance biblical authority because a deeper understanding of the workings of nature led to a deeper understanding of God.

Exploring the history of geologic thinking about the biblical flood reveals how cultural friction generated conflict and change within both scientific and religious circles. After all, the story of Noah's Flood provided the first geologic theory to be tested against field observations. Perplexing questions, like the origin of valleys and marine fossils found within mountains, became fodder for a grand debate over Noah's Flood. And arguing about evidence for a global flood likewise helped shape how biblical interpretations adapted to scientific advances. Today, unraveling the origin of traditional flood stories involves not only the interpretation of foundational Judeo-Christian traditions but understanding conflict between visionaries and orthodoxy within scientific and religious establishments.

Scientifically inclined creationists tend to be engineers, chemists, and physicists with little to no geological training. Perhaps this helps explain why the creationist view of Earth as only a few thousand years old contrasts with geology textbooks based on decades of research confirming that we live on a planet that is four and a half billion years old. Rejecting conventional geology out of hand, creationists selectively interpret the rock record to support their view that Noah's Flood deposited all the fossil-bearing rocks and sculpted the world's topography over the course of a single year. In such a short span of time a flood of epic proportions is the only geological mechanism that could do it. It's all creationists have that can explain earth history, and without it their intellectual house of cards comes crashing down.

Whatever you may think about evolution, the creationist belief in a several-thousand-year-old Earth shaped by Noah's Flood is as scientifically illiterate as the idea that the Sun circles us. Both have been known to be wrong for centuries. And to embrace the creationist view of earth history is to deny Earth's autobiography inscribed on pages of stone.

The land beneath our feet is active, changing, and moving—every day, somewhere. We simply cannot afford to ignore what we learn from geology. We use it to find oil, site and design buildings, map floodplains, and assess mineral deposits. Science is useful because it explains how the world works. This is why we place faith in it.

The history of thinking about catastrophic floods certainly features its share of conflict. Yet amid the conflict I found fertile cross-pollination between geology and Christianity. Scientific discoveries shaped creative explanations for earth history, and the interpretation of biblical stories of the Creation and Noah's Flood framed the ideas of early geologists. The challenge of adapting biblical interpretations to accommodate geological discoveries helped shape modern Christianity, influencing both liberal and conservative thought.

Let me take you on a journey through the story of how geologists learned to read the history of the world in the rocks beneath their feet and the hills above their heads. Instead of the familiar tale of controversy over Darwin's ideas, we'll see how geological discoveries helped trigger a different story of evolution—that of Christian theology and the birth of modern creationism. Along the way we'll explore how one of humanity's fundamental traits—observing the natural and physical world around us—led to stories about unimaginable floods. You see, the stories of Noah's Flood and the Tibetan flood are much the same, except of course that one went viral and we're still arguing about it. We'll also see how creationists came to consider reason in general, and geology in particular, as the enemy of faith, so much so that they could not bring themselves to accept scientific findings that seemed to corroborate biblical stories. So, like Alice heading down the rabbit hole, let's start at the beginning.

For a geologist, the logical place to begin is in the oldest rocks buried at the bottom of the geologic record. I know of no better place to see how a geologist reads a story of rocks, topography, and time than the Grand Canyon. This stunning landscape tells a tale stretching back into deep time over an unimaginably vast expanse of earth history. Armed with a few commonsense rules to guide reading the rock record, one finds in the canyon a story of whole worlds come and gone long before the one we know. The story is laid out plain as day in the walls of the deepest hole in North America.

2

A Grand Canyon

FINALLY I TOOK the last step and reached the top. It had taken all day, but I had fulfilled an ambition to hike up through the world's best-exposed story written in stone. Standing on the rim, I turned and looked down almost a mile to the bottom of the Grand Canyon, still marveling over the extraordinary tale preserved in the rock walls along the trail. Elated and exhausted, I left the rim and walked over to the National Park Service gift shop.

I picked up a small coffee-table book intriguingly titled *Grand Canyon: A Different View*. It told of how Noah's Flood ripped up the surface of the world like a geological blender, laid down the great pile of rock exposed in the canyon walls, and then deftly excavated the canyon as the waters receded.[1]

Digging deeper into the book, I read that the canyon itself was carved when the sediment that formed the rocks now exposed in its walls was still soft. I was puzzled that the authors did not try to explain how a mile-high stack of saturated sediment remained standing without slumping into the growing chasm—or how all the loose sand and clay later turned into solid rock. The book simply stated that, according to the Bible, Noah's Flood formed the Grand Canyon and all the rocks through which it's cut in under a year. There was no explanation for the multiple alternating layers of different rock types, the erosional gaps in the rock sequence that spoke of ages of lost time, or the remarkable order to the various fossils in the canyon walls. The story was nothing like the tale I read in the rocks I had spent the day hiking past.

The long plod out of the canyon still rang in my head as I returned the book to the shelf and stepped back outside. I savored the view and my day immersed in geologic time. Reading about earth history is one thing; to see and feel it for oneself is another.

I thought back to the beginning of my day, just after dawn. The towering rock walls rising above the bottom of the canyon baked in the early morning light as they've done for countless years. My knees still ached from the hike down two days ago; and the trail rising a vertical mile ahead promised another brutal hike under the Arizona sun. There was no alternative. I was committed to climbing out of one of the deepest holes in the world, passing through time from the dawn of life in the depths of the canyon to the modern desert at the top.

I approached the Colorado River, the clear turquoise water marking the start of the trail back up to the canyon rim. Watching the river flow beneath me as I crossed the footbridge, it dawned on me that the sediment-trapping Glen Canyon dam almost a hundred miles upriver robbed the river of the sand and erosive power that together cut a narrow slot into the hard rock exposed along the canyon floor.

Halfway across the river, at the far side of the bridge I saw a tunnel enter the rock wall rising from the river's edge. I entered it and felt like I'd stepped back into deep time.

In the smooth rock walls I saw the signature of abrasive sand-charged floods surging down the canyon. The surface of the hard, crystalline Vishnu Schist was a polished face made of intergrown quartz, feldspar, and mica stretched and folded at high temperature and pressure, deformed into great swirling patterns. Deep within the earth, below a now-vanished mountain range, the schist in front of me had crystallized long before dinosaurs, about a third of the way back through geologic time. But it didn't start out as hard rock. Ghost beds of sand lie preserved as light-colored, quartz-and-feldspar-rich layers sandwiched between dark layers of ancient mud now baked into aluminum-rich mica and garnet. This layering is a telltale sign that the schist formed when the sand and mud of an ancient seabed were buried deep enough to recrystallize and deform like melting ice cream.

To get hard rock to flow requires both extreme heat and high pressure. Recrystallizing and deforming the particular combination of minerals in the Vishnu Schist takes temperatures of 900–1300°F and more than three thousand times atmospheric pressure. Geologists know from temperatures measured at the bottom of deep drill holes that it gets 104–122°F (40 to 50°C) hotter with every mile below ground. We can surmise from this that the schist was approximately ten miles below the surface when it formed, twice as far down as Mt. Everest is tall. The bottom of the canyon exposes the roots of an ancient mountain range, visible today only because of the erosion of the overlying rock that had to have lain above the canyon walls to turn all that sand and mud into solid rock in the first place.

How long ago did the schist form below those ancient mountains? More than a billion years ago, although no one can tell just by looking at the rocks. Using the right tools, the age of a rock can be read like a geologic clock because radioactive isotopes decay at a fixed rate. Radiometric dating is based on the fact that younger rocks have more of the initial parent isotopes of their radioactive elements and older rocks have proportionately more of the daughter isotopes produced by radioactive decay. If you know the half-life of an isotope—

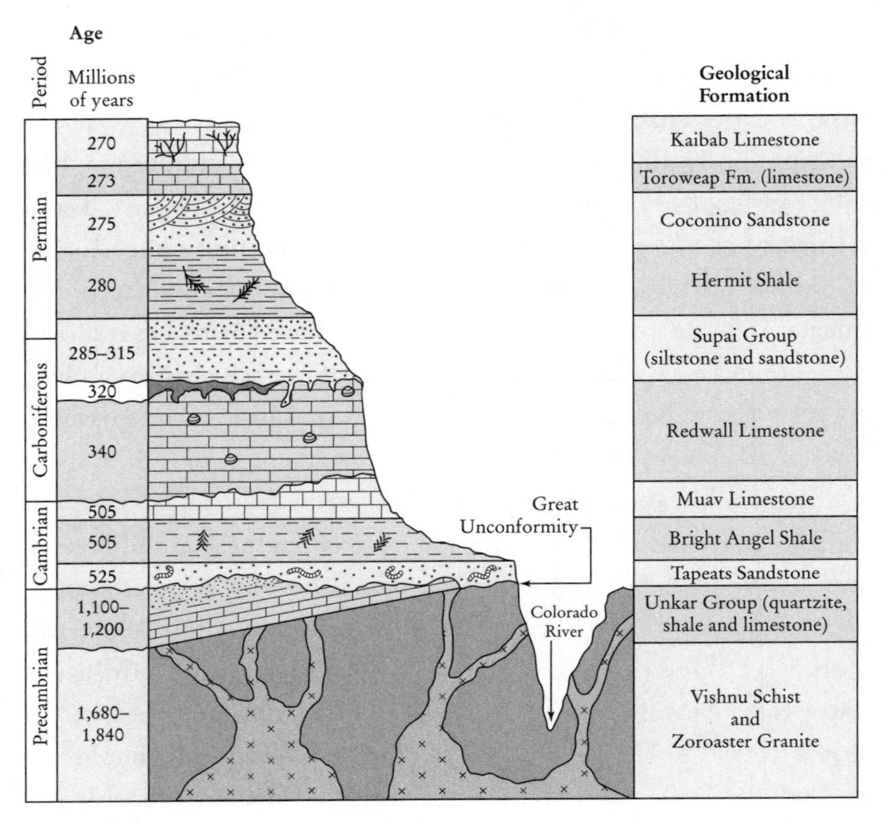

Stratigraphic column showing the rock formations exposed in the walls of the Grand Canyon (based on a sketch by Véronique Robigou).

how long it takes for half the remaining amount to decay—then the ratio of the parent-to-daughter isotope now in a rock tells you how long ago the rock crystallized.

Uranium-lead dating is the radiometric method commonly used to date the oldest rocks. The 4.47-billion-year half-life for the decay of uranium 238 (^{238}U) to lead 206 (^{206}Pb) is about the planet's age. Zircon, an uncommon mineral found in rocks like granite, is ideal for this dating method, as it strongly excludes lead from its structure upon crystallization. Because of this, all the ^{206}Pb in zircon had to have come from the decay of ^{238}U, since the mineral cooled. Geologists use a mass spectrometer to measure the amount of ^{206}Pb and ^{238}U in a grain of zircon and determine the age of the rock from the ratio of the two isotopes.

Continuing along the trail, thin bands of pink granite rise like fossilized stripes and cut through the swirling Vishnu Schist. These narrow lines of granite, called dikes, have uranium-lead ages of up to almost 1.7 billion years and neatly truncate the banding in the schist, adding a geometric flourish to the fluid forms crystallized in the inner canyon's rock wall. I could tell the schist is even older because the granite dikes cooled in cracks within it. The schist was already there when the granite cooled.

I felt like an ant crawling along the narrow trail as I snaked my way out of the inner gorge. Once I gained enough elevation, I could look down on the river and across to the other side. There, I could see a sequence of tilted rock layers on top of the Vishnu Schist and tucked in below the flat-lying Tapeats Sandstone that defined the upper lip of the inner gorge. Pitched up at a jaunty ten- or twelve-degree slant, this stack of limestone, shale, and quartzite (a hard rock made when sandstone is heated deep within the earth) records the changing depth of water in an ancient sea, deep-water limestone on the bottom giving way first to shale made from offshore mud and then to sandstone from a fossilized shoreline.

Whether eroded by rivers, wind, or waves, the truncated upper surface of the schist, together with the hardened marine sediment sitting right on top of it, is evidence that rocks once buried miles underground were brought to the surface, exposed to the elements, and then buried again deep below the bed of an ancient sea. After all this, the whole package got tilted up and planed off by erosion for a second time before being capped by the sediment composing the still flat-lying rocks rising far above. I could see it all laid out in the cliff, right across the river—two rounds of uplift and erosion buried beneath a three-thousand-foot-high wall of rock. An ancient story only unveiled because the Colorado River carved the Grand Canyon.

The striking outcrops of the inner gorge illustrate several simple rules geologists have used for centuries to read the story of rocks the world over. The first is that layers of sedimentary rock that accumulate in depositional environments, like sandstone and shale, represent time and are deposited horizontally. This makes sense, since these rocks form by settling through water under the influence of gravity, the way mud settles to the bottom of a glass of water. The second is that rocks at the bottom of a pile are older than those above them. This, too, seems obvious. And where one formation cuts off another, it is the younger that cuts across the older. Using these simple rules—determining what's above what, and what cuts through or across what else—is how geologists decipher stories of time and change written in stone. Of course, there is more to knowing how to read Earth's story, such as how two other types of rocks—igneous and metamorphic—are made, but simple interpretive rules apply to all rock types.[2]

Everywhere on Earth is either eroding and losing material or receiving deposits of material eroded off of somewhere else—one geologic realm sheds sediment, the other accumulates it. But the places where each is happening change over time. The most obvious change apparent in the walls of the Grand Canyon is that the marine rocks exposed in it have switched from one domain (deposition) to the other (erosion). Eroded upland environments are not preserved in the rock record because there's nothing left to see—they've vanished. The geologic signature of mountains is recorded by its absence, a gap in the record of time, while the story of our planet and life on it is archived in the sediments of depositional lowlands and marine environments—the places where sediment piles up over geologic time.

Deciphering earth history involves establishing the basic relationships between different rock formations and the nature of the boundaries, or contacts, between them. Two layers of sedimentary rock deposited one atop the other without any discontinuity are considered conformable—they accumulated with minimal interruption. An eroded surface leaves a discontinuity between two rock units, a gap representing missing time that geologists call an unconformity. An unconformity represents how far down erosion wore into an ancient landscape before additional sediment was deposited on top. A whole series of unconformities exposed in the canyon walls tell of multiple rounds of deposition, deformation, and erosion before the whole package of rocks rose from the sea to the level at which we find them eroding today.

After far too many switchbacks, I made my way out of the inner canyon and across a cliff of flat-lying rock. In passing, I traced my finger along the surface of the unconformity where the Tapeats Sandstone rests on the irregular surface of the Vishnu Schist. The now

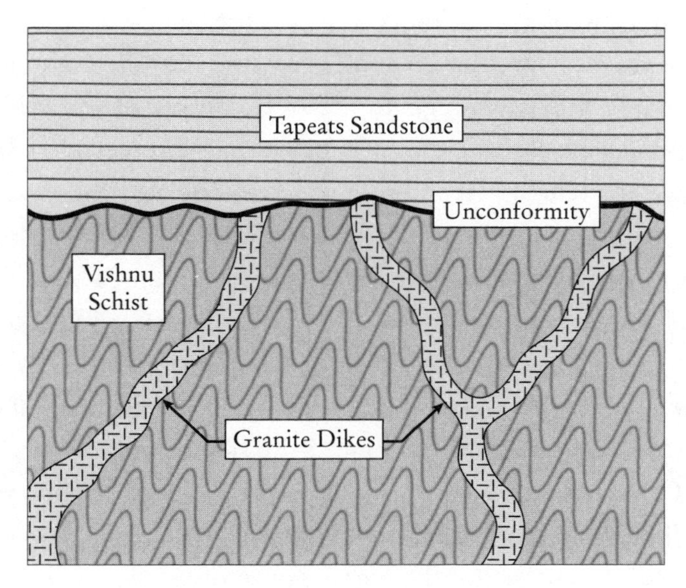

The great unconformity at the base of the Grand Canyon where the Tapeats Sandstone truncates the Vishnu Schist (based on a sketch by Véronique Robigou).

solidified grains of sand settled onto a rocky seabed in the Cambrian Period, about 100 million years before plants began colonizing land.

The span of time this unconformity represents is staggering. More time lies missing between the 1.7-billion-year-old Vishnu Schist and the 525-million-year-old Tapeats Sandstone than is recorded in the enormous wall of rock soaring thousands of feet overhead. All the lost time was enough to erase a mountain range and hide the ruins of ancient worlds, familiar in design but alien in detail. My imagination wrestled with how a thousand million years could vanish from the geologic record. Two worlds had come and gone, leaving nothing behind but their rocky bones.

I kept walking up, leaving the unconformity behind, and noticed burrows and tracks of simple wormlike animals in the cliff of Tapeats Sandstone through which the trail ran. The fact that marine life was crawling around the bottom of an ancient sea, as documented in the solid evidence right in front of me, presents a serious problem for the creationist view of the Grand Canyon. How could fragile worms have been crawling around on and burrowing into the seafloor during a flood powerful enough to remodel the planet? The biblical flood would have to have dumped more than ten feet of sediment every day for a whole year in order to have deposited the thousands of feet of sediment exposed in the canyon walls.

Finally I reached an area known as Indian Gardens, a broad bench a third of the way up the canyon. This scrap of flat ground offered shade and a welcome break from slogging up switchbacks. Trailside exposures of the hard-baked mud of the Bright Angel Shale showed why this relatively flat ground graced the side of the canyon. Rock fragments burst from beneath my boot as I kicked an outcrop in passing. The shale was too weak to hold a cliff.

I looked closely at the shale and saw more signs of life in the tracks and burrows preserved on paper-thin layers of stone. Here was proof of an ancient seabed made from slowly accumulating mud in which animals crawled, foraged, and died. People more patient or lucky than I in inspecting these rocks have found trilobites, distant cousins of spiders and crabs that roamed the seas from 515 million years ago to when they perished in the Permian Period, almost 200 million years before the demise of the dinosaurs.

Pausing on the gentle terrain of Indian Gardens offered me a panorama of grand vistas. Yet the far rim lay out of sight. As I surveyed the staggering expanse of the canyon, it was hard to fathom how long it would take to erode the cliff at the canyon's edge back beyond the horizon, one rockfall at a time.

Lingering in the tree-shaded oasis of Indian Gardens, I reflected on how people see the land in different ways. More than rocks and topography, beliefs and experiences have shaped attempts to explain the formation of the Grand Canyon. People living along the canyon were curious about how it formed long before geologists and creationists formed their views.

Like the Tibetan villagers along the Tsangpo, Native Americans living along the Colorado River have stories about a giant flood. A Havasupai story attributes the carving of the Grand Canyon to a catastrophic flood that occurred when the mischievous god Ho-ko-ma-ta unleashed a tremendous rainstorm. Another, more thoughtful god, Pu-keh-eh, put his daughter in a hollowed-out log to save her from the monstrous current rushing down to the sea. After the floodwaters receded, she crawled out from her improvised vessel and became the mother of all humanity. The story shares the same broad outline as that of Noah's Flood—only with a matriarchal figure.

Still soaking up the geologic story, I began climbing again toward the canyon rim. At the far end of Indian Gardens the shale gave way to the Muav Limestone, a section of carbonate rock sitting directly on top of the Bright Angel Shale. The sequence of sandstone, shale, and limestone I had just walked told me that the water deepened over time. At the upper boundary of the Muav Limestone, another unconformity lies at the base of a formidable cliff. This gap in the rock record documents another hundred million years lost to erosion above sea level.

What you find directly on top of the Muav Limestone depends on your location in the canyon. From where I was, the trail snaked up a looming wall of red-stained rock rising several hundred feet upward. Climbing through a slot in the dark cliff, the trail followed a fractured zone in the aptly named Redwall Limestone. Along the trail I spotted clamlike fossils, evidence that life in the rocks was starting to become more visible, more complex, and more familiar. Yet another unconformity lies atop the Redwall, this one perforated by caves and sinkholes that formed when percolating rainwater dissolved soluble limestone and made a Swiss cheese–like landscape. This unconformity represents another 25 million years between the deposition of the Redwall Limestone and the overlying rocks of the Supai Group.

Once above the cliff, the trail crossed back and forth countless times from siltstone to sandstone to siltstone and back to sandstone again. I was climbing a geological staircase made of shale treads and sandstone risers. Each step recorded the rise and fall of an ancient sea, with harder sandstone forming cliffs, and weaker shale forming gentle ledges. This means that the ancient sediments turned to stone before the canyon carved down into them. If the now solidified rock had been wet and loose when the canyon was cut, the canyon walls would reflect the strength of loose sediment rather than that of the rock. The sandstone would not hold cliffs because loose sand can only support slopes of at most 30 to 40 degrees, as you can see for yourself in the produce section of any grocery store by pulling an orange from the bottom of the pile. And that's when it's dry. Saturated sand can hold a slope only about half that steep—nowhere near a vertical wall. In contrast, wet clay is cohesive enough to support short vertical cliffs. If the creationist view held water—that the canyon's slopes formed when the sediment was still saturated—then the slopes today would have shale cliffs and sandstone benches, the opposite of what's visible along the trail.

How long does it take for the finest sediment to settle out? Even in a bucket of still water it can take weeks for fine clay to drop to the bottom. The distinctive microrhythm of coarse to fine, coarse to fine, coarse to fine in the walls of the canyon proves how the now rock-solid sediment settled out from a series of flows. The hundreds of thousands, if not millions, of layers of silt could not have settled out and separated from the intervening layers of sand during the passage of a single violent current because turbulence would have resuspended the fine sediment. Individual layers of clay, silt, and sand take a long time to segregate out—and far longer to do so over and over again to build up a pile it takes hours to hike up through.

Every step brought me closer to the canyon rim, and I passed into the soft, easily eroded Hermit Shale, where the slope relaxes and piñon pines and juniper trees manage to hold down a thin soil. Within the rust-colored shale below the soil, fossil ferns, conifers, and the tracks of reptiles and amphibians revealed the former nature of the region. I had climbed out of an ancient ocean and into the remains of a temperate coastal jungle.

Passing out of the Hermit Shale, I started up more switchbacks and crossed onto a massive, strikingly white sandstone. Composed of pure quartz sand, the Coconino Sandstone exhibits cross-beds that define the faces of fossilized sand dunes rising diagonally through cliffs along the trail. Invertebrate tracks and burrows are preserved in these fossil dunes. One outcrop near the trail preserves reptilian footprints that displaced the sand, sending it slumping back down the face of a dune. Such fine-scale features would have been obliterated if they had formed underwater, the way waves running up a beach erase footprints in the sand. These dunes were made by wind.

Continuing my trudge, I passed through the yellow-gray Toroweap Formation, evidence that a sea submerged the desert sands of the underlying Coconino Sandstone. Then a final, grueling climb up the nearly vertical white wall of Kaibab Limestone. Plodding past day hikers heading down the trail, I could see 270-million-year-old fossil coral and mollusks that reminded me of the complexity and diversity of life I'd seen learning to scuba dive on Australia's Great Barrier Reef. Finally, I reached the top.

Standing on the rim, I reviewed the story I read hiking up through ancient worlds—missing mountains, shallow and deepwater seas, coastal jungles, windswept sand dunes, and coral reefs. The few simple organisms in the lowest layers offered an obvious contrast to the complex reef community at the top. I had completed a grand tour through geologic time telling of the rise and fall of ancient mountains and seas, with the rocks at the bottom reaching back to the dawn of life and those at the top predating the dinosaurs. That the cliffs were solid rock right up to the canyon rim testified to erosion of all the formerly overlying rock that provided enough pressure to solidify a pile of sand and mud in the first place. Whole worlds came and went before the one we know today.

One doesn't need to hike up through thousands of feet of rock outcrops to refute the notion that the Grand Canyon formed during a flood that somehow managed to lay down all those rocks right before carving through them to create the canyon. A simple experiment you can conduct at home will prove the point. Get a glass-walled box (a fish tank will do nicely), fill it with water, and pour in a mix of clay powder, sand, and pebbles. Larger particles and denser material will settle out first, forming a pile with pebbles on the bottom, sand in the middle, and clay on top. Then, pour in rocks or sand all of the same grain size but a mix of colors, and you'll get a collage for a deposit. In order to sort by color, you have to add one color at a time. The rocks exposed in the walls of the Grand Canyon could not have settled out during a single flood because they alternate many times in color, grain size, and composition.

Something that really struck me about my hike up the canyon was how the plants and animals entombed in the walls of the canyon are extinct. If all the creatures buried in that mile-high wall of rock had been put there by the biblical flood, then why aren't modern animals entombed among them? That the vast majority of fossils are extinct species presents a fundamental problem for anyone trying to argue that fossils were deposited by a flood from which Noah saved a pair of every living thing.

A simpler, fatal problem for the creationist interpretation of Grand Canyon geology is that sandstone, such as the cliff-forming Tapeats and Coconino formations I hiked past, and shale, like the Bright Angel formation I kicked apart so easily, form under completely different conditions than limestones like the Muav, Redwall, Toroweap, and Kaibab. Marine limestone forms when organisms whose bodies are made of calcium carbonate ($CaCO_3$)—like coral, clams, or microscopic foraminifera—die. Their shells and skeletons pile up on the seafloor and, if subjected to enough pressure, temperature, and time, eventually form carbonate rock. Because the organisms that become carbonate rocks take time to grow and don't live in turbid waters, the alternating layers of biologically precipitated limestone and mechanically deposited sandstone and shale that settled out from turbid water could not have formed during the same event. The alternating sequence of different rock types stacked one atop the other in the canyon walls records a long series of events and environments.

A single enormous flood simply can't explain the geology of the Grand Canyon. As we'll see, geologists discredited the idea that Noah's Flood created the world's topography and deposited its sedimentary rocks in the early nineteenth century, decades before the first expedition down the Grand Canyon. When creationists argue that they want to "teach both sides" of the argument about earth history in science classes—their view and those of the "evolutionists" they vilify—they neglect to mention that geologists had disproved the creationist view of a young Earth shaped by Noah's Flood before Darwin ever began thinking about evolution.

Today, the real debate about the formation of the Grand Canyon is between geologists who agree about its geologic history but argue about its topographic history—exactly how and when the canyon itself formed. The conventional view is that by six million years ago the Colorado River had established its modern course to the Gulf of California. Radiometric dates on cave deposits record that the water table draining into the river dropped steadily downward at about a hundredth of an inch a year for at least the past three million years. Younger lava flows that spilled into the canyon also tell of gradual incision over the past half-million years. But a recently published study suggests that the modern canyon was preceded by an older canyon first carved by a river that drained in the opposite direction between eighty and seventy million years ago. Arguments about the topographic evolution of the canyon center on whether the modern canyon formed as a river eroded headward back into the Colorado Plateau, a plateau-bound lake spilled over a drainage divide to cut a new path to the Gulf of Mexico, or local uplift pushed rocks up around a river with enough erosive power to maintain its course.

Although a great flood did not carve the canyon itself, there is evidence of grand floods within it. Breaching of cooled lava dams that impounded the river may have launched catastrophic floods down through the canyon. One of these natural dams was over two thousand feet tall. Flood deposits found within the canyon include huge boulders perched hundreds of feet above the river. No doubt a flood capable of stranding boulders so high on the canyon walls would have been spectacular—had anyone been around to see them. But most of these floods occurred long before people made it to the

New World. Native American tales of how the canyon formed are attempts to make sense of mysterious landforms.

In contrast to how simply and directly creationist claims of a global flood can be refuted, it took centuries to compile the rough outlines of earth history. Generations of geologists competed to find key outcrops, develop new theories, and demolish the ideas of intellectual rivals. As the world became better mapped, several revolutions tied it all together—the discovery of geologic time, recognition of how both gradual and catastrophic action carve topography, and the revelation of how plate tectonics shapes Earth's surface. Geologists today are confident about reconstructing earth history because geological mapping and correlation, radiometric dating, and fossils all tell a consistent story. Geology provides an independently verifiable answer to the age-old question of how the world we know came to be.

All people, geologists included, tell stories to explain the world around them and thereby understand our place in it. Different ways to see the Grand Canyon led to very different interpretations of the canyon and what it means. It might seem reasonable to think that a global flood carved the canyon if all you have to explain earth history is what you read in the Bible. But if you let the rocks speak for themselves they tell another story, just as grand: about the unimaginable depths of geologic time instead of the devastating power of a single flood.

How did it come to be that today what one sees along the Bright Angel Trail contradicts what many consider a Christian view of the canyon? To answer this question, we need to explore a two-thousand-year-old running argument about how to interpret nature and the story of Noah's Flood. Perhaps the best place to start is at the top—of the world.

3

Bones in the Mountains

Consider Mount Everest. The world's highest mountain consists of three geological formations separated by two faults, shattered zones across which rock formations slid into place. Much as the layer-cake rock sequence exposed in the Grand Canyon captures the scope of geologic time, the shuffled geology of Mount Everest reveals the power of unimaginably slow deformation to transform the bottom of the sea into three different kinds of rocks and stack them back up to crown the world. This would have been unimaginable to early Christians. Climb up the mountain and you can see it for yourself.

After leaving Katmandu and trekking more than a week through the glacier-carved valley of the Dodh Kosi river, you'd arrive at Everest base camp, 17,590 feet above sea level. From there it is another eleven and a half thousand feet up to the top. The bottom half of the mountain, the part below about 23,000 feet, consists of the Rongbuk Formation, metamorphic rock with a composition similar to granite. Like the Vishnu Schist at the bottom of the Grand Canyon, the

Rongbuk Formation formed when marine sediments were buried miles below ground.

The suite of minerals in the Rongbuk indicates it formed at temperatures of 1000–1250°F and at 8,000–10,000 times atmospheric pressure, more than fifteen miles down in Earth's crust. Radiometric ages of unaltered mineral inclusions in the Rongbuk reveal that the original marine sediment was deposited some 490 million years ago. Once the rocks were stacked up into enough of a pile to heat up, deform, and start melting its own base, numerous granitic dikes rose like crystallized tendrils climbing their way up toward the surface.

Continuing up through shattered rock to cross the Lhotse Detachment, the lower of two fault zones, you'd reach the North Col Formation, which extends up to about 28,200 feet. This formation consists of 490-million-year-old marble, schist, and phyllite—limestone, sand, and mud buried deep enough to be pressure-cooked into harder rocks, but not so deep as to start melting. The mineral assemblage in the North Col Formation shows it underwent metamorphism just two to four miles below ground, at temperatures of 850–950°F and pressures of 1,000–2,000 times atmospheric. It was never buried anywhere near as deep as the rock right below it. Missing are the miles of rock that must have once lain between the now neighboring rock formations.

At the top of the North Col Formation, a distinctive stripe of yellowish marble (metamorphosed limestone) called the Yellow Band cuts across the mountain. At the top of the Yellow Band, a zone of completely shattered rock defines the second fault zone, the Qomolangma Fault, which separates the marble below from unmodified limestone of the overlying Qomolangma (or Everest) Formation. These uppermost rocks also date from about 490 million years ago, and extend to the summit 29,035 feet above sea level.[1] The three rocks of Everest were born in the same sea, but they had radically different histories before being spliced together to form the world's highest mountain.

Standing on the frigid summit of Everest, if you could pick up a piece of the limestone and view it under a microscope you would find that the top of the world consists of fragmented trilobites and tiny fecal pellets that settled to a tropical seabed. Beneath your boots you'd see the essential truth of the world's highest mountain—the rock at its top once lay at the bottom of the sea.

How could a scrap of seafloor come to cap the world? Based on the cooling history of minerals they contain, these rocks started rising from the sea about fifty million years ago, when India began smashing into Asia. As India moved north, Asia stayed put, crumpling, folding, and faulting the incoming rock that had been deposited in a shallow sea. Crushed in a geological vise, the old seafloor was squeezed up and up, rising centimeters a year to eventually stand more than five miles above the coast. Faults formed as the incoming rock compressed, fractured, and pushed aside the rock that was already there. The southern edge of the Tibetan Plateau began to slide down toward India in much the same way that material starts sloughing off the top and sides of a pile of sand if a bulldozer keeps advancing into it.

But if you didn't know about plate tectonics, how could you explain finding an old ocean floor on top of the planet's highest peak? People around the world faced a similar question when they saw marine fossils entombed in high mountains. One way to resolve such puzzles is to assume that mountains don't rise and that an incredibly deep sea once covered the peak, and thus the whole world. Another way is to assume that the rocks now exposed in the mountain somehow rose miles up out of the sea. Imagining that Noah's Flood submerged the Himalaya is no less intuitive than the modern scientific idea that India is slamming into Asia and bulldozing up the world's highest mountains in a process so slow one could not observe its progress over many lifetimes.

If you think the world is static, the idea of deforming and deconstructing rocks into whole new formations would never cross your mind. Before the concept of geologic time entered into people's thinking, it was crazy to imagine that India was pushing up an old seabed to form the Himalaya. Faced with the choice between a catastrophic flood or mysteriously rising mountains, early natural philosophers considered a mammoth flood less preposterous.

Naturally, arguments erupted about how to interpret and reconcile religious beliefs with discoveries about nature, and vice versa. How could they not? Humanity's essential curiosity and propensity to talk promote debate. Was Genesis intended as a concise history of the Jewish people, a literal and comprehensive history of the world, or as metaphorical parables for ages to come? The modern creationist concept of fundamental conflict between faith and reason would have shocked early Christians who believed that discoveries about the world revealed natural truths that could only support biblical truths.

Noah's Flood was a powerful narrative that greatly affected the early development of geology because natural philosophers initially looked to the biblical flood to explain rocks, topography, and whole landscapes. How could the shells of sea creatures come to rest inside mountains? Discoveries of marine fossils found far above the sea bolstered the view of Noah's Flood as a global catastrophe. The idea that the world had been reshaped by a great flood doubled as biblical truth and the first geological theory for much of postclassical antiquity.

In ancient Greece, however, there was a wide range of strikingly modern ideas about why mountains contained marine fossils. Some of the earliest known philosophers recognized the organic nature of fossils as creatures that lived in a remote time long before people walked the Earth. Fossil seashells told of oceans that covered the land. Giant vertebrae and enormous teeth that were occasionally unearthed were widely recognized as ancient bones. Fossils discovered near sites of legendary battles were displayed in temples as the remains of epic heroes or mythical monsters. The Greek idea that modern animals and people were but puny shadows of bygone days reinforced the widespread belief that the world was running down, wearing out, and growing old.

One might even be tempted to consider the great philosopher Aristotle a protogeologist for recognizing that landscapes evolved over unimaginably long time spans. In his view, land and sea constantly swapped places, and marine fossils in the rocks of mountains testified to how sea could become land. Rivers carried silt and sand to the sea, gradually filling it in, causing the sea level to rise and submerge coastal areas. This endless cycle in which land became sea and then land again so slowly as to escape observation paralleled Aristotle's belief in a world without beginning or end. Civilizations rose and fell before they could record even a single round of this grand cycle. The world was eternal and always changing.

Philo offered one of the earliest surviving commentaries on Noah's Flood in his *Questions and Answers on Genesis*, published in the first century AD. Born into an aristocratic Jewish family in Greek-ruled Alexandria, Philo didn't question the historical veracity of the biblical flood. He was primarily interested in revealing the true meaning of scriptural passages. To him, this meant exploring deeper, allegorical meanings. He considered literal interpretations superficial. Philo singlehandedly initiated both sides of a long history of novel and conflicting interpretations of Noah's Flood. He characterized the biblical flood as both limitless, having drowned the whole earth, and as having flowed almost beyond Gibraltar, implying that its influence was restricted to the Mediterranean.

Whether he meant to or not, Philo articulated both sides of what would become a grand debate among generations of theologians and natural philosophers. Did the biblical flood inundate the entire planet or just the world known to Noah? Christians debated this question long before science entered the fray. At stake was how to evaluate the truth about the world. Do you have faith in what you already think you know, or do you adapt your thinking to new information? Ever since, this question has been at the heart of an ongoing conversation between faith and reason. And the story of Noah's Flood has put these different styles of belief into direct conflict perhaps more than any scientific issue other than evolution.

Among those arguing about how to read Noah's story was Celsus, a second-century Greek philosopher. An opponent of Christianity, he charged the Jews with borrowing Noah's story from pagan sources. Biblical critics like Celsus questioned the ability of the ark to hold pairs of all the world's animals. How could one build such a boat? The preposterous story of a farmer building a lifeboat for all of creation seemed like a Jewish fairy tale.

In response, the second-century church father Origen countered that Genesis should be understood figuratively.

> *Now what man of intelligence will believe that the first and second and the third day, and the evening and the morning existed without sun and moon and stars. . . . I do not think anyone will doubt that these are figurative expressions which indicate certain mysteries through a semblance of history and not through actual events.*[2]

Origen invoked Greek culture in promoting a figurative reading of the story of Noah. Why did his contemporaries allow Greek myths allegorical meanings but insist on literal meanings for the biblical story? To him, the symbolic meaning of Noah's Flood was as important as its historicity. Noah foreshadowed Christ, the animals stood for the kingdom of Christ, and the ark represented the church—the ark's three decks symbolized heaven, Earth, and the underworld. In his mind, a literal reading did not do Noah's story justice.

Origen's insistence on allegorical readings was not unique. Christians in his era tended to interpret biblical stories allegorically to encourage moral behavior. Sensitive to pagan critiques like those of Celsus, Christian philosophers advocated using knowledge of the natural world to better understand the Bible. Clement, Origen's teacher and head of the Catechetical School in Alexandria, chided those who did not wish to use logic and reason in interpreting the holy book. He embraced both faith and reason. Understanding the truth expressed in God's creation could only lead to a better understanding of God. Clement held that Christians should bring all knowledge to bear on the truth because the world could not contradict its creator. To him the bond between faith and reason was as close as that between God and Christ.

Saint Augustine stands out among early Christians who wrestled with such questions. Born in Roman Africa in 354 AD to a pagan father and a Christian mother, Augustine was educated in Carthage, where he became familiar with classical knowledge, Latin literature, and pagan beliefs. A brilliant intellect who lived a hedonistic lifestyle as a youth, he rose to become professor of rhetoric at the imperial court in Milan, the most visible academic post of his day. His worldly experiences before converting to Christianity in his early thirties helped frame an attitude of belief in what one could see firsthand. In his view, nature didn't lie. He interpreted fossil shells and bones entombed in the fabric of the land as natural evidence that verified the story of Noah's Flood.

Remarkable for the clarity of his thoughts about the relationship between rational and spiritual life, Augustine warned of the danger in embracing biblical interpretations that conflicted with reason. Fearing that Christians could lose faith when confronted by evidence contradicting sanctioned interpretations of scripture, Augustine wrote:

> *Let no one think that, because the Psalmist says, He established the earth above the water, we must use this testimony of Holy Scripture*

> *against these people who engage in learned discussions. . . . Ignorant of the sense of these words, they will more readily scorn our sacred books than disavow the knowledge they have acquired by unassailable arguments or proved by the evidence of experience.*[3]

Secure in his faith that Scripture and the natural world shared a common author, Augustine advocated flexible biblical interpretation that could be adjusted in light of what one learned about the natural world. He advised Christians to avoid endorsing biblical interpretations contradicted by what they could see for themselves.

Augustine also defended the idea that Noah's Flood covered the whole planet by employing explanations based on the knowledge of his day. When critics argued that floodwaters could not have risen higher than the lighter clouds surrounding Mount Olympus, Augustine countered that Olympus itself towered over the clouds despite being made of earth, the heaviest element. Why, therefore, could not water rise as high for a brief time? While this argument seems rather silly today, it sounded rational at the time and shows Augustine's flexible thinking in reasoning about the nature of the world. To him, one could make sense of natural and physical phenomena so long as one had a keen eye and a curious mind.

To Augustine, the most compelling evidence for a global flood was the widespread occurrence of plant and animal remains in rocks. Fossils seemed to tell the story as plainly as the Bible. Far more interesting and controversial were questions about the symbolic meanings and significance of Noah's story.

Augustine's contemporary, Saint Jerome, translated the Bible into Latin and institutionalized allegorical interpretations. Jerome also extolled the virtues of thoughtful reasoning in understanding scripture. Holding Earth's disrupted, broken, and twisted crust as evidence of God's wrath, he considered literal interpretation of the Bible as shallow reasoning. Jerome cemented within the church a tradition of considering literal interpretations for the illiterate masses and allegory for more advanced minds—that is, the clergy. For a thousand years it was the clergy's job to offer deeper and more meaningful interpretations for those lacking the interest, commitment, or intellect to take on the task. Eventually, the tide shifted when Martin Luther led the sixteenth-century Protestant rebellion against an elite, allegorically minded priesthood, reclaiming the banner of biblical interpretation for the more literal-minded.

Jerome's translation of Genesis introduced unintended fodder for conflicting interpretations when he chose to translate the Hebrew word "adamah" to Latin as *terra*, "earth," instead of *humus*, "soil." His choice of earth instead of soil for this passage (Genesis 3:17) in the Latin Bible sparked debate about whether God cursed the whole planet or just the fields tilled by man. If earth meant soil, then Adam's punishment consisted of having to work the land for a living. But if God cursed Earth itself, then perhaps topography was a manifestation of divine vengeance, the lasting signature of a world-shattering catastrophe. This (mis)translation would greatly influence fellow Christians who believed in the ongoing degeneration of both humanity and the world following Adam and Eve's fall from grace.

Both Jewish and early Christian traditions held that mountains formed after God created the world, which initially was a more perfect form, like a sphere or an egg. Some held that God scooped out the ocean basins and piled up the spoils to form continents and mountains a couple of days before he created people. Others thought that topography arose from sin but argued over the timing. Perhaps God inflicted the inconvenience of mountains to punish Adam and Eve when they were expelled from the Garden of Eden. Or maybe mountains formed when He cursed Earth for receiving Abel's blood. Many of those who pondered such things believed that topography formed when Noah's Flood reworked Earth's surface. Whether formed before or during the Flood, the irregular form of mountains testified to how God could extend his punishment of humanity to scarring the face of a once perfect Paradise.

In this vein, early Christians generally considered fossil seashells relics of Noah's Flood, tangible reminders of humanity's depravity. Through the Middle Ages Christian theologians taught that the ongoing decay of the world mirrored mankind's spiritual and moral degeneration. Where today we see high mountains and dramatic landforms as iconic natural cathedrals embodying the wonder of creation, for centuries the Christian perspective was just the opposite.

Augustine's views endured in those of thirteenth-century Catholic philosopher Saint Thomas Aquinas. Like Augustine, he advocated flexibility in interpreting Genesis. He thought that because the church was eternal, Christianity could wait until natural philosophers determined what was certain before deciding which of the possible interpretations of Genesis to abandon in the face of apparent contradictions. Although Aquinas accepted the reality of Noah's Flood, he promoted understanding the book of nature—God's other book—in seeking to understand both scripture and the world around us. God created reason and endowed humanity with the ability to judge truth and the free will to embrace or deny it. Aquinas allowed no room for conflict between the Creator and how the world worked. He considered such conflict a logical absurdity.

Aquinas and Augustine viewed reason as a fruitful gift and a way for people to embrace and practice learning about things larger and more meaningful than one's self. To me, this sounds perfectly consistent with how geologists like myself, and scientists in disciplines from astronomy to zoology, conduct our inquiries. I didn't expect to find the bedrock principle underlying science enshrined in early Christian thought.

Still, times have changed. In Aquinas's day, three generally accepted facts about earth history were rooted in the teachings of the church. The world was a few thousand years old, Noah's Flood reshaped topography, and everything would end in a great conflagration at the end of the millennium (although, as we'll see, opinions differed as to just when that would be).

Later in the Renaissance, the rediscovery and translation of influential Greek and Arabic philosophical texts blurred the distinction between living and nonliving things. If Earth itself was alive, perhaps fossils, a name that covered any odd thing found in a rock, could grow in rocks. Stalactites dripping from the ceiling of caves grew within the earth. Why not fossils too? Such thinking led natural philosophers to see fossils as objects that simply mimicked the shapes of living organisms. While natural philosophers came to regard fossils as nothing more than mineral curiosities, a few, like Leonardo da Vinci, thought otherwise.

Late in the fifteenth century, the rivers and hills of northern Italy fascinated the son of a public official in the town of Vinci, nestled at the foot of Monte Albano. As a boy Leonardo wandered up the mountain and found a cave where the rock walls were a hash of seashells and fish bones. A natural skeptic, he didn't believe the common explanation that Noah's Flood had carried the shells into the mountains. His doubts were strengthened when, years later, he worked on canal projects where excavations exposed numerous fossils embedded in solid rock. Observing his surroundings, Leonardo concluded that a great flood did not entomb marine life in stone. Some shells were clamped shut, as if buried alive. Others were broken into fragments and scattered in deposits resembling modern beaches. The surfaces of rock layers even preserved worm tracks. He may have been the first to question whether worms could crawl around the seafloor and leave perfectly shaped, undisturbed tracks during an epic flood.

Watching how flowing water moves sediment, Leonardo concluded that no flood could have carried ancient seashells into the mountains for the simple reason that fossils and other objects heavier than water sank to the bottom of a current. Fossils were neither souvenirs of the Flood nor inanimate curiosities. Either God was trying to trick him, or the story was more complicated than implied by a simple reading of Genesis.

Leonardo reasoned that layers of sedimentary rock initially formed from mud that gradually settled to the bottom of an ancient sea. Fossil shells preserved in the rocks high on ridges were deposited during an era of higher sea level. Trusting reason and the testimony of his own eyes to decipher the structure of God's grand design, he saw no evidence of a catastrophic deluge.

Even if Noah's Flood had drowned the world, Leonardo did not see how it could have carved topography. If it rained enough to submerge the highest peaks, the floodwaters would have formed a great sphere. But were water to everywhere rise to the same elevation, it would have no slope to propel it. How could the floodwaters erode valleys without moving? Besides, where did all that water go afterward? For a mind such as Leonardo's, more looking and thinking only spawned more questions.

Getting rid of the floodwaters presented as great a challenge as generating a global flood. Evaporating a globe-covering mass of water would require more heat than the Sun could muster. And not only were shells heavy enough to settle out in turbulent water, but the water at the bottom of a wave moves away from shore. Noah's Flood would have dragged fossils out to sea rather than pushed them up onto mountains. To Leonardo, fossil shells entombed in upland rocks, the conventional evidence for a global flood, amounted to no evidence at all.

Later, exploration of the New World would raise new problems for a global flood. Particularly troublesome was the huge increase in the number of species Noah had to house on his ark as explorers discovered the world's great variety of life-forms. As confounding as how all of these new animals could have fit aboard was the question of how they traveled to the ark before the flood and then back home again afterward, all without leaving any offspring in the Old World.

Unlike Leonardo, who stuck close to home, everywhere European explorers went they found people who didn't appear to be descended from a Jewish patriarch. Biblical apologists proposed that Native Americans descended from the lost tribes of Israel, from Viking expeditions, or from people who had crossed ancient land bridges to the New World. Such solutions introduced even more problems. Where were these continent-connecting land bridges now? Could Pygmies, Vikings, and Aborigines all have descended from Noah in just a few thousand years, when classical statues revealed that Greeks and Italians looked the same two thousand years ago as they do today? If people changed so slowly, how could the kaleidoscope of the world's ethnicities have developed since Noah's Flood? However one looked at it, the biblical account provided an incomplete view of earth history.

The discussion changed with the arrival of Protestant thought. The reformers who split the church broke with the centuries-long Catholic tradition of allegorical interpretation but could not agree among themselves about how to read the story of Noah's Flood. Protestants introduced both more literal and liberal interpretations as they taught all people to interpret the Bible for themselves.

Unlike their contemporaries in the sixteenth-century Catholic church, Martin Luther and John Calvin ignored the implications of New World discoveries. They were religious reformers, not explorers faced with conundrums manifest in the flesh of exotic animals and peoples. But here again we find more debate than uniformity of thought. The two great minds that laid the intellectual foundation of the Protestant church, and all its denominational offspring, offered opposing interpretations of Noah's Flood. In their commentaries we can recognize a resemblance to scientific rivals hashing out how to interpret puzzling data.

Published in 1545, Luther's *Lectures on Genesis* devoted more than a hundred pages to commentary on Noah's Flood. He declared that Moses "spoke properly and plainly, and neither allegorically nor figuratively."[4] He held that the Flood annihilated the earthly paradise and left no trace of Earth's original surface in its wake. Petrified wood and fossils dug out of mines, the buried ruins of the former world, were all that was left to testify to the destruction of humanity's cradle. Generating the Flood was no problem because God held the continents above the seas through divine buoyancy He could rescind on command.

And then, like the coat of a dog shaking off after a bath, the surface of the world went from flat to wrinkled. A quick dunk and shake sums up how Luther's Flood reshaped the world to create modern topography. Some areas rose to become mountains. Others sank beneath the seas. The Flood destroyed Earth's original soil that had produced incredible bounty with little labor. "Before the Flood turnips were better than melons, oranges, or pomegranates were afterwards."[5] Luther even asserted that the Flood began in springtime to maximize the terror for a populace "full of the expectation of a new year."[6] Clearly, such opinions expand upon a literal interpretation of Genesis, if only because, like dinosaurs, turnips are not mentioned anywhere in the Bible. Given his propensity to supply details of his own, even Luther, someone generally considered a strict biblical literalist, struggled with biblical interpretation.

Having grown up in the tamed, rolling hills of lowland Germany, Luther was unaccustomed to and intimidated by alpine topography. To his eye, the ragged nature of mountains mirrored mankind's spiritual deterioration. Mankind had been in decline since the chaos of the Flood resurfaced the world and left mountains tarnishing the face of creation.

Luther's fellow reformer John Calvin also endorsed a literal interpretation of the biblical flood but did not fill in the kind of detail that

Luther offered up. Noting a lack of consensus on such matters, Calvin did not offer fossils as evidence of a global flood. In contrast to Luther, he maintained that after the Flood the world remained in roughly its former state. Rather than a catastrophic reshuffling of the physical world, Calvin's version of Noah's Flood served as a quiet reset button.

Unlike Luther, Calvin lived much of his life in and around the Swiss Alps. He loved nature and could not believe God would create a world that was not beautifully rugged. Neither could he believe that God would curse the world itself on account of humanity's sins. Just as reason elevated men above beasts, nature was a lens through which to behold God. And if Earth did not share in God's curse, then how could mountains have been created during Noah's Flood?

These two traditions that trace back to the roots of the Protestant church essentially stake out different ways of dealing with the relationship between science and religion. The Protestant followers of Calvin encouraged study of the natural world in seeking to understand the universe and humanity's role in it, an approach paralleled in the Jesuit tradition of Catholic scholarship in natural philosophy. While Calvin's accommodating views fostered a spirit of scientific inquiry, Luther's cultivation of more literal followers led to a less flexible understanding of the natural world. Although the two great reformers differed on how to interpret Noah's Flood, they both thought Nicolaus Copernicus heretical to challenge the conventional view that the Sun circled us.

Copernicus announced his radical theory that we circled the Sun as a visiting scholar in Rome around 1500. At first he cast the idea as an intellectual curiosity, a novelty to exercise the mind. Later, after decades contemplating the matter, he became convinced that this was indeed how the world worked. And although Pope Clement VII reacted favorably to the idea in the gardens of the Vatican, Copernicus returned to his hometown in Poland rather than tangle with the papal censors in Rome when he dedicated his *On the Revolutions of the Heavenly Spheres* to Pope Paul III in 1543. Unbeknownst to him, his publisher added a groveling preface that apologized for ideas intended as hypothetical speculation rather than fact. An anguished Copernicus only learned of this duplicity on his deathbed when he first glimpsed his just-published book.

Copernicus was not the only one disappointed with his book. Ever the literalist, Luther was appalled by the suggestion that our world was not the center of the universe. His plain-sense understanding of scripture led him to denounce such egregious heresy. "This fool wishes to reverse the entire science of astronomy; but sacred Scripture tells us that Joshua commanded the sun to stand still, and not the earth."[7] The ideas that Jerusalem was the center of the world and that Earth was the center of the universe were solidly enshrined in Christian doctrine. Besides, the classical theory that the Sun circled Earth seemed to account for the movement of heavenly bodies. How else could Joshua have commanded the Sun to stand still (Joshua 10:12–13)? Over the next several centuries, Calvin's attitude of greater flexibility in how to interpret natural phenomena helped generations of Protestants accept scientific revelations.

Half a century later, Galileo Galilei inadvertently supported Copernicus and tested another Pope's patience by pointing his newly invented telescope at Jupiter in 1610. His discovery that moons circled another planet took Copernicus's hypothesis out of the realm of speculation. If moons orbited other planets, then might not Earth itself orbit the Sun? Although he prudently named Jupiter's moons after his Medici patrons, Galileo was still denounced as an enemy of Christian faith.

Scholars eager to defend the Bible agreed that Galileo's findings were absurd. When he offered doubters a chance look through his telescope, many either proclaimed it impious to look or denounced Jupiter's tiny satellites as devilish illusions.

Turning his telescope toward the Moon, Galileo made another

heretical discovery—plainly visible mountains. This was a problem, for mountains were not supposed to be there. If Earth's topography resulted from Noah's Flood or Adam's Fall, then why would similar features scar the surface of the Moon? It made no sense for man's curse to extend to worlds where no sinners lived.

This time Galileo had gone too far. His support for the Copernican system was labeled atheistic, and he was denounced to the Inquisition in Rome.

Attempting to defuse the controversy, Galileo wrote to his friend Grand Duchess Christina of Lorraine and argued that literal interpretations of the Bible should not be applied to scientific questions. His critics were missing the point and needed to think more liberally.

> *Contrary to the sense of the Bible and the intention of the holy Fathers . . . they would have us altogether abandon reason and the evidence of our senses in favor of some biblical passage, though under the surface meaning of its words this passage may contain a different sense.*[8]

Galileo further argued that the study of nature reveals facts about the way the world works—but that the Bible is notoriously difficult to interpret.

> *If anyone shall set the authority of Holy Writ against clear and manifest reason, he who does this knows not what he has undertaken: for he opposes to the truth not the meaning of the Bible, which is beyond his comprehension, but rather his own interpretation; not what is in the Bible, but what he has found in himself and imagines to be there.*[9]

Galileo was saying that the problem lay in how one read scripture rather than in anything one could observe and study about the world. To his way of thinking, apparent conflicts between scripture and reason could be resolved if one reinterpreted the Bible on the basis of careful observation of nature, on the basis of natural facts. New discoveries could guide biblical interpretation on matters pertaining to the natural world.

Galileo further defended Copernican theory and his own thinking by arguing that Moses adapted his language to his audience. Today one generally does not try to teach quantum physics in high school, or James Joyce to the illiterate. You can't teach someone something he or she lacks the background to learn.

Although the Inquisition could not condemn Galileo for observing something, interpreting scripture was a different matter. The Council of Trent had forbidden interpretations that contradicted the traditional commonsense views of the church fathers. And an Earth-centered universe was enshrined in Catholic tradition. To argue otherwise was heresy.

When informed of Galileo's correspondence in 1615, the Inquisition convened a handpicked panel of theologians who were ordered to judge propositions extracted from his letters. They obediently ruled that "the proposition that the sun is the centre and does not revolve about the earth, is foolish, absurd, false in theology, and heretical, because expressly contrary to Holy Scripture."[10] In February of the next year, Pope Paul V ordered Galileo brought before the Inquisition, where Cardinal Bellarmin decried the damage it would do to Christian faith were the planets found to revolve around the Sun. If Earth was nothing special, just one of many planets careening through space, how special were its inhabitants? Galileo's telescope not only threatened humanity's favored place in the eyes of God, it threatened the Bible's promise of salvation.

Galileo found himself in ever more awkward quarters. How could one individual challenge the most powerful political and cultural force of his day? In his own defense, Galileo invoked the authority of St. Augustine's ideas, but even that didn't work.

Several weeks later the Inquisition condemned an already dead

Copernicus and banned all writing that affirmed that Earth revolved around the Sun. To teach that our planet moved through space was dangerous in this world and invited damnation in the next.

After Pope Urban VIII permitted Galileo to write a book outlining the arguments for and against the Copernican system, Galileo eventually published his *Dialogue Concerning the Two Chief World Systems* in 1632. The price of publication was the condition that Galileo include the pope's views and yet another humiliating preface admitting that Copernicus had fabricated it all. This time, however, scholars all across the continent laughed at the transparently coerced disclaimer. If Galileo secretly felt redeemed, it did him no good. He didn't help himself by putting the pope's traditional views in the mouth of a character named Simplicio, which can be interpreted as simpleton. The embarrassed and infuriated Pope ordered Galileo to his knees in front of a tribunal and forced him to recant his heretical ideas.

Galileo's experience shows how conflict arose when science revealed things that contradicted traditional beliefs. It also raised a still controversial question: How were Christians supposed to react to the discoveries of natural philosophers? Did empirical observation trump biblical revelation, or vice versa? That this issue remains unresolved is apparent in the arguments used in today's ongoing conflict over what to teach in science classrooms.

Although Galileo endured clerical condemnation for arguing that Earth was not the center of the universe, the then conventional idea that Earth stood at the center of everything came from the Greek geographer Ptolemy. The Bible does not directly address the issue. Neither does it address the date of creation. The belief that the Bible says we live on a not-quite six-thousand-year-old Earth at the center of the universe is itself an interpretation. Gradually, the idea that there were other ways to interpret biblical stories came to be accepted. By the time Pope John Paul II apologized publicly for Galileo's persecution in 1992, the church had long since abandoned the idea of Noah's Flood as a global deluge. The new official view was that those who condemned Galileo did not recognize the potential for differing interpretations of the Bible's plain words.

Consider, for example, how through a literal interpretation one can read something into the Bible one knows not to be true, like that the world is flat. The Creation story in Genesis says Earth is covered by a great vault (firmament) on which the celestial bodies move across the sky, which makes literal sense if the world is flat—like the floor of a grand temple. And must not Daniel have considered Earth essentially flat when he interpreted the dream of a great tree that could be seen to the farthest end of the world (Daniel 4:20)? This only would be possible if the world were flat (and a lot smaller than it actually is). Obviously, it is impossible to see the far side of the world on a spherical planet, which is why one understands the obvious meaning as a figure of speech.

This is not just an Old Testament problem. Literal interpretation of the New Testament also implies a flat Earth. Matthew wrote that the devil showed Jesus all the kingdoms of the world from the top of a high mountain (Matthew 4:8). This would only be possible if the planet were indeed flat, unless of course Matthew was referring to all the kingdoms of the Middle East, the world known to the Jews. Similarly, the Book of Revelation refers to "the four corners of the earth" (Revelation 7:1) despite the fact that spheres lack corners. In other words, acknowledging the fact that we live on a planet requires allowing for figurative or allegorical interpretations for these, and therefore other, biblical passages.

As debate about the nature of the cosmos, the beginning of the world, and evidence for the Flood moved from cloisters into more public forums, Protestants generally promoted biblical literalism in their feud with the Catholic Church and its allegorical readings of the Bible. Today, however, few realize that until the Reformation

Christian theologians considered strict biblical literalism simplistic fodder for the illiterate masses.

Questioning traditional biblical ideas about the natural world became less dangerous in the decades after Galileo's ordeal. Despite substantial friction between religious denominations (not to mention a few wars), natural philosophers investigating Earth and the cosmos developed experimental approaches to scientific inquiry and proposed imaginative theories to rationally explain Noah's Flood through secondary, natural causes rather than miracles. Although science as we know it was yet to emerge, scholars increasingly believed that investigating the natural world held the key to deciphering the mysteries of God's creation. Observation paved the way to insight. Those investigating nature were confident that they would not only confirm the truth of a global flood but discover how cleverly God pulled it off—and reveal just what the Bible meant in describing how "all the fountains of the great abyss were released, and the floodgates of heaven were opened" (Genesis 7:11).

The history of attempts to understand the Bible shows that what one reads into it can be as influential as what it says. As people learned more about the world, certainty in the reality of Noah's Flood led to imaginative ideas for reconciling geological evidence with biblical stories. But instead of resolving the issue, these efforts created new divisions, because the harder people looked for evidence of a global deluge, the less convincing the case for one became.

4

World in Ruins

Long before geology developed into a distinct discipline in the nineteenth century, novel theories abounded about how Noah's Flood shaped the world. In Galileo's day, three camps defined seventeenth-century views of topography. First, those who did not think too deeply about such things generally believed landscapes were just two days younger than Earth itself, sculpted by the hand of God on the third day of creation. Then, there was the more scholarly view that valleys and mountains were carved by the Flood. Finally, some natural philosophers allowed the earthquakes known to have happened occasionally through history a minor role in shaping the land. Conventional wisdom still held that the world had been gradually wearing down through its short history. The future promised further decay as topography eroded and soils lost fertility.

The view of the world as a wrecked and ruined place began to change in 1644 when renowned philosopher René Descartes set forth how Noah's Flood followed his principles of nature—the laws of physics as he laid them out in his *Principia Philosophiae*. One of his theories concerned Earth's origin and evolution. Mindful of Galileo's treatment by the church, and well aware that his ideas did not accord with sanctioned interpretations, Descartes explicitly stated that his own theory was wrong. Cleverly inoculated from official censure, he claimed to offer a hypothesis useful for better contemplating nature.

Descartes painted a picture of an Earth that began as a failed star trapped in the vortex of a neighboring star. The primitive Earth then cooled and segregated into a planet with distinct layers, leaving a still fiery core surrounded by a metal-rich inner crust. Above this lay an ocean, trapped below an outer crust made of stones, sand, and clay. Over time the heat of the Sun evaporated water trapped between the inner and outer crusts. Fissures coalesced into large fractures as the undermined and weakened outer crust foundered into the watery abyss, triggering a great flood and forming both mountains and seas.

Descartes' imaginative idea offered a way to generate the world's topography all at once. His grand physical explanation for how to generate a global flood inspired other natural philosophers to think up ways to trigger the biblical flood. With little evidence available to contradict or refute any idea no matter how outrageous, competing flood theories soon posed creative ways to explain how God designed a world preprogrammed for destruction.

Today, such theories seem fantastically ridiculous, like bizarre figments of feverish minds. But in their day, they were serious attempts to explain the world. Imagination raced ahead of understanding as the reality of Noah's Flood was taken on faith in theories devised to explain the origin of topography. Facts only started to get in the way of a good theory once geological principles were systematized.

After Galileo's ordeal, Jesuit scholar Athanasius Kircher became a leading voice among clergy interested in natural history. Professor of mathematics, physics, and Oriental languages at the Jesuit College of Rome, he published lavishly illustrated natural history books that became wildly popular among the European elite. An eccentric by any standard, Kircher explored deep grottoes and canyons, even having himself lowered into the volcanic craters of Etna and Vesuvius to see what lay below ground. Finding subterranean streams high in the Alps, he saw the fact that some caves were filled with water and others with fire as the key to one of Earth's great mysteries—the origin of rivers. His *Mundus Subterraneus* (*Subterranean World*), an encyclopedic compilation of geologic fact and fable published in 1664, suggested that ocean tides pumped seawater up into mountains through underground channels that connected to springs at the head of rivers. Fires deep beneath volcanoes, acting like a global radiator system, drove water up from holes in the bottom of the sea to feed mountain springs. Kircher had the concept of a hydrological cycle right, but the direction backwards. Today we know that water evaporates from the oceans and rains down on the continents and then runs off into the sea.

A decade later, in his *Arca Noë* (*Noah's Ark*), Kircher maintained that God unleashed Noah's Flood by causing vast underground lakes to overflow. Great blocks of the planet's outer shell foundered into his subterranean reservoirs, leaving distorted layers of broken rock standing above ocean basins and lowlands. Mountains were the collapsed ruins of Earth's original crust.

Not everyone was convinced the flood was global. Kircher's contemporary Isaac Vossius, Dutch theologian and librarian to the Queen of Sweden, argued for a local flood on the grounds that there simply was not enough water on Earth to submerge the highest mountains. He dismissed as pious fooleries proposals that God miraculously created extra water and then just as miraculously made it all disappear. Vossius argued that the few generations between Adam and the Flood could hardly have populated Mesopotamia, let alone the entire planet. Instead, he proposed that people must have occupied a limited area in Noah's time because it was senseless for

God to punish uninhabited places. Besides, the ancients often used universal terms to describe local events. The Flood need only have been universal in the sense that it overwhelmed humanity's ancestral homeland. In his reading, the Bible revealed Noah's Flood to have been a local affair.

The amount of water required to flood the world also was a sticking point for Edward Stillingfleet, the Anglican Bishop of Worcester, who in 1666 wrote *Origines Sacrae* (*Sacred Origins*). He too considered a local flood consistent with biblical orthodoxy. According to his calculations, the world's clouds could only produce enough water to cover the globe with a foot and a half of water—nowhere near enough to submerge the whole planet. Stillingfleet echoed Vossius in thinking that a regional flood could have destroyed mankind if humanity was restricted to the Middle East. A flood that affected a small part of the world would also mean that Noah only needed to load representatives of part of the animal kingdom on his ark. Stillingfleet did not favor invoking additional miracles not mentioned in scripture to explain a worldwide flood, or the logistical challenge of feeding a boatload of animals when all the world's edible plants lay submerged beneath the waves.

Stillingfleet and Vossius helped establish the legitimacy of belief in a local flood among theologians, but the propensity to interpret Noah's Flood as a global deluge did not fade easily. Prominent seventeenth-century natural philosophers continued to use Noah's Flood to explain geological observations, among them the grandfather of geology.

The Dane Niels Stensen, better known as Steno, was the son of a successful Copenhagen goldsmith. Born into a Lutheran family on New Year's Day in 1638, Steno was taught that at most the world would last another couple of centuries before God ended everything. His deep religious faith and strong interest in natural philosophy greatly influenced how he came to lay the foundation for modern geology. Raised in a Protestant stronghold of biblical literalism, he later worked and lived in Catholic countries where allegorical interpretations of the Bible were deeply rooted. His gradual migration south would change his worldview and encourage his curious, wondering mind to think broadly.

At the age of eighteen, Steno enrolled at the University of Copenhagen to study medicine. There he learned the supposed curative properties and medicinal virtues of crystals and fossils such as tongue stones—rock-hard triangular objects with serrated edges. Prized since ancient times, powdered tongue stones were thought to ward off evil or attract affection and were commonly sold as cures for plague and bad breath. They could be found scattered on bare ground after heavy rainstorms, and there were many theories of how they formed. Some thought the strange objects fell from the sky. Others thought that they were petrified lightning strikes. While tongue stones and fossils interested Steno, he loved anatomy lessons involving the dissection of human bodies.

In 1659 Steno slipped out of Copenhagen, eluding the Swedish troops besieging the city. After a brief stay in Amsterdam, he finished his medical training at the University of Leiden. There his skill as an anatomist led him to the scientific discovery that made him famous. Recreationally dissecting a sheep's head, he discovered the saliva duct. Until then how saliva got to the mouth was a mystery. He went on to discover tear glands, disproving the conventional wisdom that pain or grief squeezed tears from the brain.

Following his graduation in the winter of 1665, Steno came to Paris. There, he boldly challenged Descartes' claim that the tiny pineal gland housed the human soul near the seat of the brain. Steno's careful dissection of human brains disproved the great philosopher's assertion that the nut-shaped gland twisted and pulled strings animating the human body. Steno showed that the pineal gland was held fast and could not gyrate. He continued to startle the scientific establishment in Paris when he then contributed significantly to understanding the workings of the human heart.

Now a scientific sensation, Steno was offered the position of physician to Ferdinand II, Grand Duke of Tuscany. With this came access to the Accademia del Cimento (Academy of Experiment), the first and only formal research lab of its day—founded by students of Galileo and supported by the grand duke's deep pockets. Steno's journey to Florence carried him across the Alps and Apennines, where he saw fossils layered in rocks high above sea level, well beyond the reach of even the largest waves. Some rock layers lay flat, others were contorted and lay at steep angles. While the fossils in the hills around Florence looked like seashells, most natural philosophers did not consider them signs of ancient life. The educated consensus was that they were insignificant mineral oddities, sports of nature that merely resembled oysters and clams.

Soon after Steno arrived, in October 1666, fishermen on the Tuscan coast hauled in the body of a monstrous great white shark near the mouth of the Arno River. When word of the several-ton beast reached the Medici palace, Ferdinand ordered it brought to his court in Florence for the Accademia to examine. But the shark was too large to transport and was already starting to rot. So its enormous head, as big as a whole pig, was loaded onto a horse-drawn cart and sent up the Arno River valley.

Steno, the academy's newest member, considered the honor of dissecting the enormous shark's head a once-in-a-lifetime opportunity. He cut as the grand duke and a mesmerized crowd of courtiers watched. The jaws were large enough to swallow a man whole. Yet its brain was tiny—just three ounces. How could such a diminutive brain control a giant killing machine?

Steno focused first on its teeth. Each serrated blade was identical to the mysterious tongue stones. They were as identical "as one egg resembles another."[1] Seeing that tongue stones were actually shark's teeth, he wondered how the teeth of giant sharks could end up enclosed in solid rock. They must have become fossilized after laying in the mud of an ancient seabed that somehow became stranded high above the sea.

Steno described his findings in a short report to the grand duke, with a digression on the origin of tongue stones and the implications for understanding other fossils. He pointed out the flaw in the conventional wisdom of the time: that fossils spontaneously grew within rocks. A growing object would crack the rock, yet one never saw cracks around fossils found in rocks. Even more telling was that tongue stones were always perfect replicas of their biological counterparts. In contrast, most crystals contained a defect, even when grown in a lab. Steno argued that fossils resembling broken mussel shells found with their matching halves preserved in rock inches away from each other could only be explained as the remains of once living creatures.

Steno's demonstration that tongue stones were petrified shark teeth convinced scholars that fossils were indeed organic remains. His interest in the problem of solids enclosed within solids—how fossils got into rocks—led Steno to deduce that the bottom layers in a pile of sediment were deposited first. This is the foundational principle of modern geology, Steno's Law of Superposition—the idea that the oldest sedimentary layers are on the bottom and the youngest are on top. It's still valid centuries later; I used this same basic rule to interpret the geologic story when I hiked out of the Grand Canyon.

Steno thought that some rocks were made of consolidated sediment washed off the land and that other rocks precipitated from mineral laden waters. Fossils were the remains of sea creatures buried by gradual deposition of sediment on the seabed. This was why fossils tended to be the most durable parts of marine creatures (teeth, bones, and shells). Soft tissue decayed too rapidly to be preserved.

Steno's prescience is astounding given the time in which he lived and the countervailing convictions of his peers. The impact of his shark head dissection and the short yet wildly influential publication it spawned in the spring of 1667 show the serendipitous nature of scientific progress. Steno subsequently began working on a longer masterpiece that laid geology's foundation. In trying to explain how shark's teeth ended up in rocks, he devised rules for how to read geologic history from the rocks themselves. Whereas Descartes and Kircher developed their ideas from sweeping generalities based on classical ideas backed up by little, if any, geologic evidence, Steno studied Earth's history by applying guiding principles and logic. He didn't just make up a good story to explain how he thought things worked; he went out and scoured the countryside for clues to build up ideas that were grounded in field evidence.

As he grew increasingly enamored with geological problems, Steno began collecting fossils on long hikes in the Tuscan mountains. Indulging Steno's curiosity, the grand duke opened quarries and mines to expose what lay underground. The more Steno observed, the more he became convinced that an ancient sea deposited fossil-bearing rocks. He also noted how some rock layers lay at an angle to the horizon, meaning that they had been tipped up on end after they were deposited.

In his hallmark contribution to geology, Steno adopted guiding principles for interpreting the history of rocks that he intended to be so simple, clear, and transparent that no one could dispute them. The first simply states that the layers at the bottom of a pile of sediment were laid down first, and therefore were oldest. The second holds that sedimentary layers are deposited horizontally. These simple principles allowed him to start piecing together the story of the Tuscan landscape. Defining how to determine the relative age of strata and past events opened the door to deciphering earth history. It was Steno's distinction between primary and secondary rocks—between crystalline rocks he thought were made at the initial Creation and layered rocks that formed later from detritus eroded off the original rocks—that set the stage for the development of a geological time scale.

By the spring of 1668, Steno's trips into the hills convinced him that the ancients were right about the nature of fossils. That summer he submitted his findings about fossils and Tuscan geology to the church's censors, who routinely vetted the theological acceptability of scholarly discoveries, opinions, and interpretations about the natural world. Although this meant his *Dissertation on Solids Naturally Enclosed in Solids* wasn't published until the following year, Steno did not need to worry about meeting Galileo's fate. His interpretation of geological evidence as faithfully recording the biblical flood placated the church even though he broke with prior tradition to interpret earth history through studying rocks and fossils rather than scripture.

In contrast to the fanciful theories of his better-known contemporaries, Steno's interpretation of how Noah's Flood shaped the Tuscan landscape was rooted in field observations. After laying out how fossils got into rocks, he described the sequence of events he read in the hills around Florence.

He concluded there were six periods in earth history that corresponded to the biblical account. He found no fossils in the lowest, and therefore oldest, layers. So these rocks formed right after the Creation, when water covered the world. Before the creation of life, sedimentary rocks lacking fossils settled out in this primeval sea, laid down as horizontal strata. As these newly deposited rocks emerged to form dry land, Steno thought subterranean fire or water ate out huge caverns in the underlying rock. When these great caves collapsed to produce valleys, it triggered Noah's Flood as the seas rushed down to fill the new lowlands. After more sediment settled in the new, lower-elevation sea the whole process repeated, resulting in a second round of cavern collapse that produced modern topography and the tipped-up rock layers within. It was Steno's way of making

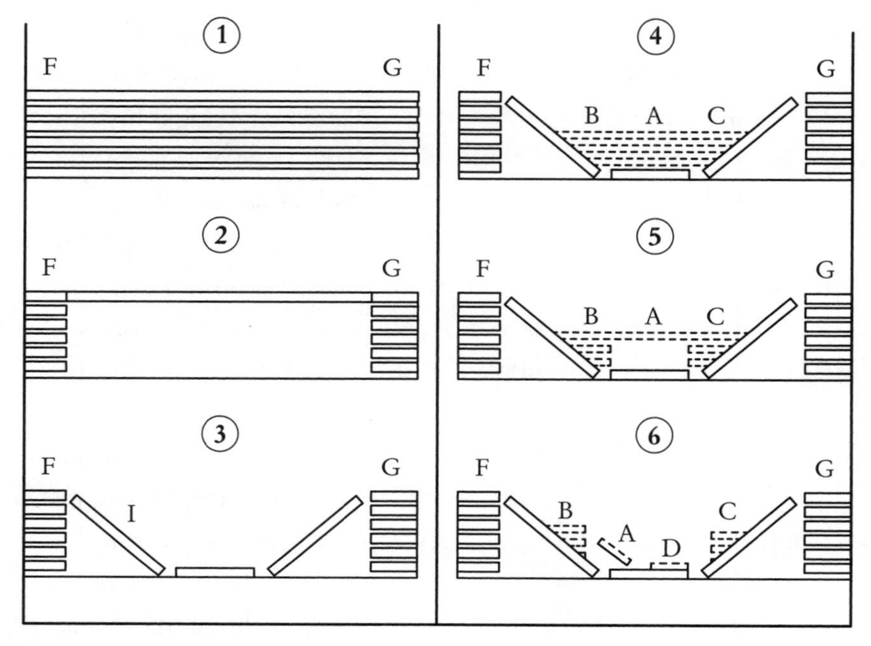

Steno's six-stage model for the formation of the landscape around Florence, involving: (1) precipitation of fossil-free sedimentary rocks into a universal ocean; (2) excavation by fire or water of great subterranean caverns beneath Earth's pristine surface; (3) collapse of undermined continents to produce a great flood (Noah's Flood); (4) deposition of new layered (sedimentary) rocks containing fossils in inundated valleys; (5) continued undermining of younger rocks in valleys; and (6) another round of collapse to create the modern topography.

sense of what he observed given his faith in the historical reality of Noah's Flood.

Steno's version of Tuscan geologic history fit neatly into the traditional interpretation of Genesis as historical truth. But where did the floodwaters come from? Steno's theory combined various natural causes to explain Noah's Flood. Debris from Earth's collapsed outer crust blocked the passages that pushed seawater back up into the mountains. While the sea overflowed, rain fell incessantly. Rivers dumped eroded soil into the sea, making it overflow all the more. In other words, Steno gathered the floodwaters from everywhere he could—sea, sky, and the subterranean abyss. While much of what he deduced about how rocks form and how fossils become part of them stood the test of time, his interpretation of the geologic history of the Tuscan landscape did not.

This would not have bothered him. Steno viewed science as a spiritual endeavor, a quest for better understanding God and better interpreting scripture. Seeing humility as important in both science and religion, he eventually became disillusioned with his colleagues' petty rivalries, arrogance, and lust for fame and converted to Catholicism on All Souls' Day, November 2, 1667.

For months he had been agonizing over whether to abandon his native faith and join the Catholic Church, troubled by the problem that Protestants and Catholics alike were convinced that theirs was the one true faith. Both could not be right. The dichotomy of these two worldviews haunted Steno and eventually led to a life-changing decision. The root of his angst was the inclination of Protestants toward literal interpretation of the Bible versus the allegorical and metaphorical lens Catholics used to address obscure passages and internal inconsistencies. And which version of the Bible was authoritative—Hebrew, Greek, or Latin? Not trusting standard translations, Steno applied his analytic powers to compare the theological claims of Protestants and Catholics against original Hebrew and Greek manuscripts in the Medici library.

In the end, however, it was not rigorous scholarship that convinced him to convert but a chance event while walking down a Florence street. Meditating on the issue, he heard a woman from an open window call for him to cross over to the other side. In addition to the implied warning about what she was about to toss out onto the street below, he interpreted this as a sign from God.

He converted, became a priest, took a vow of poverty, and gave up his studies as a sacrifice to God. Equal parts genius and saint, he routinely gave his money to the poor and often went without food himself, sometimes by choice or because he was too broke to buy it. He annoyed wealthy parishioners and fellow clergy by vociferously advocating for the poor, even selling his bishop's ring to help feed the hungry. When Steno died in 1686 his worldly possessions consisted of a few worn-out garments. Three centuries later, in 1988, Pope John Paul II beatified him, on October 23, the day that Bishop Ussher, a contemporary of Steno's whom we'll meet later, famously claimed as Earth's birthday.

Although Steno recognized some of the challenges that his geologic principles presented to conventional interpretations of scripture, he did not see the potential for fundamental conflict between science and Christianity. Like most of his peers, he thought reason helped illuminate the wonder of Creation.

Read by few in his own time, Steno's ideas caught on only after they were tested and popularized by other natural philosophers eager to use formal principles to enhance their understanding of God's greatest work, Earth. Ironically for a man of deeply conventional faith, the foundational principles he developed to explain the role of Noah's Flood in shaping the Tuscan landscape eventually undermined both the traditional biblical timeline for the Creation and the reality of a global flood.

Although geologists recognize Steno as both a pivotal and an inspirational figure, we tend to overlook the extent to which he interpreted geologic features as evidence that Noah's Flood reshaped the world. Instead, we emphasize his faith in how rocks tell the story of the world, while Steno himself was striving to read God's other book—nature. After Steno, rocks could tell their own story. The natural world and how it worked could frame—and conceivably limit or refute—theological options for how to read the biblical stories of the Creation and Noah's Flood.

Although Steno faded into obscurity in his own day, his principles have stood the test of time. In contrast, the prominent Anglican reverend and Cambridge theologian Thomas Burnet, who also worked to square geological features with Noah's Flood, left a different legacy. He proposed the most influential seventeenth-century explanation of Noah's Flood but is remembered today for the way his faith in how he read the Bible spawned an overly imaginative theory. In 1681, his elegant *Sacred Theory of the Earth* sought to address two familiar problems: where did the floodwaters come from and how did mountains form?

At Cambridge, Burnet was taught that Moses intended for the masses to interpret Genesis literally and for elite priests to read between the lines. Burnet took this view to heart when, at the age of thirty-five, he set off across Europe on a grand tour. Coming from the verdant English countryside, he was shocked by the disorder he saw in the mountains. He thought the Alps were towering wrecks composed of "wild, vast and, indigested heaps of Stones and Earth."[2]

The confusing internal structure of the Alps forced Burnet into a theological crisis. Believing that God made all things in beauty and proportion, he found the Alps a chaotic place lacking order or

Illustration of the deformed interior structure of the Alps visible in the pattern of rock outcroppings (by Alan Witschonke based on lowermost panel of plate XLVI of Johann Scheuchzer's Sacred Physics *(1731)).*

design. He could not believe God's divine hand would create such monstrous forms. Surely the Creator would make a beautiful, symmetrical world—something more like Burnet's England. Mountains must be the remains of a wrecked planet, crumbling ruins of an originally perfect sphere. Just what had happened?

After three years of travel, Burnet returned to England committed to determining how God set up a perfect world destined to disintegrate. He carefully estimated both the amount of water in the oceans and how much it would take to submerge the highest mountains. There was nowhere near enough water on the planet. It would take eight times the volume of the world's oceans. Even forty days and nights of rainfall at the astounding rate of two inches an hour would only amount to 160 feet of water. Burnet refused to invoke miracles. To assert that God simply made more water when and where He needed it was too easy, to "*cut the Knot when we cannot loose it.*"[3] There had to be another source.

Then it hit him. The world was completely different before the Flood. The water in today's oceans would cover a smooth, topography-free globe to a depth of about the biblically proscribed fifteen cubits (just under twenty-five feet). And if Earth was originally featureless, there was no ocean aboveground. Instead, a primordial ocean must have existed underground.

In Burnet's view of the Creation, God commanded the elements to sort out by density, the heaviest sinking to the center and lighter stuff forming outer layers. Gravity then settled the original chaotic mass into a dense core surrounded by a layer of water and an outer envelope of air. A greasy floating layer eventually coalesced into an outer crust like the shell of an egg. In this way, disorganized chaos became a habitable planet. Close to a perfect sphere, the primitive Earth had the perfect shape for a perfect paradise, something worthy of divine creation.

Burnet's early Earth also enjoyed an endless summer. And while this may have sounded like paradise to an Englishman, all that sunshine gradually warmed the planet, causing it to expand and form fissures at the base of the crust. It also began to dry and crack the planet's outer shell. As the great subterranean ocean heated up, its expanding vapors pressed against the planet's weakened crust.

Burnet's imagination ran with the idea as he proposed that divinely timed cracks propagated to the surface right at the peak of human wickedness, just after Noah finished building his ark. When the outer crust collapsed into the interior sea, humanity's ancestral paradise foundered into the abyss. Water shot high into the air and sloshed around for months, creating and sculpting topography. This left the planet in ruins, rugged mountains replacing smooth plains as tumultuous waves resurfaced the world.

To get rid of all the water, Burnet simply had it drain back down into cracks in the seafloor. Deep caves and volcanoes proved the presence of cavities beneath the continents. Shouldn't there be similar caverns beneath the seas? A subterranean drain also provided a handy explanation for how the seas never overflowed even though rivers drained continuously into them (although evaporation, of course, turned out to provide a better explanation).

Taking his cue from natural philosophers, Burnet did not invoke divine intervention to explain Noah's Flood. He called upon divine planning. Using reason to explain the origin of both the world and the modern landscape, he used scripture to confirm rather than define his story. "We are not to suppose that any truth concerning the Natural World can be an Enemy to Religion; for Truth cannot be an Enemy to Truth, God is not divided against himself."[4] Burnet considered it a sign of divine providence that the world was set up to trigger a flood at just the right time. His bold theory was widely hailed as a philosophical triumph.

Burnet sent an advance copy of his book to Isaac Newton, soliciting his comments. In reply, Newton cautioned that Genesis shouldn't be interpreted literally and that Moses described reality in terms understandable to the common man. Newton even proposed an unusual theory of his own to explain how hills and mountains might have precipitated out of a chaotic primordial fluid: "Milk is as uniform a liquor as the chaos was. If beer be poured into it, and the mixture let stand till it be dry, the surface of the curdled substance will appear as rugged and mountainous as the earth in any place."[5] Newton was particularly troubled by how, according to Burnet's theory, the oceans did not exist until after the Flood. If so, fish and other marine life could not have been made at the Creation. This would have required a second round of creation not mentioned in the Bible. And that was unthinkable.

Burnet's grand theory had more unorthodox implications. In particular, the problem of how Noah's descendants came to populate America after the Flood was difficult to reconcile with Burnet's broken planet. In contrast, it was easy to explain how they made it to America before the Flood—they walked. So he proposed that although Native Americans were descended from Adam, Columbus was the first of Noah's progeny to reach America. Like Noah, a few people survived the Flood on other continents. Backed into this awkward claim, Burnet abandoned literal interpretation of scripture to save his theory, which was based on just such an interpretation.

Despite the problems with Burnet's theorizing, his *Sacred Theory of the Earth* attracted so much attention that King William III had it translated from Latin into English, bringing accolades and opportunities Burnet's way. Appointed chaplain to the king, Burnet seemed sure to become primate of the Anglican Church. But the Church of England forced him into early retirement when he rashly suggested that the Fall and the days of Creation were meant allegorically rather than literally.

Critics were quick to point out that since there could be no ocean on Burnet's smooth pre-Flood Earth, there should be no marine fossils in rocks that formed before Noah's Flood. Marine fossils should only be found in younger post-Flood deposits. Yet such fossils were widely distributed through the rocks that Burnet claimed formed as part of Earth's original shell, and only later fell into the subterranean sea. Were Burnet right, this could not be.

And how did sea creatures come to exist without a second round of Creation if the oceans formed during the Deluge? Could Adam have been given dominion over the fish in the sea in an oceanless world? Herbert Croft, the aging Bishop of Hereford, labeled Burnet's theory a work of "extravagant fancies and vain fopperies" and speculated that perhaps "his Brain is crakt with over-love of his own Invention."[6]

Although he saw his theory as consistent with biblical teaching, Burnet was not simply trying to reconcile faith and reason. He was trying to prove that reason offered an independent source of revelation coequal to and compatible with scripture.

> *'Tis a dangerous thing to ingage the authority of Scripture in disputes about the Natural World, in opposition to Reason; lest Time, which brings all things to light, should discover that to be evidently false which we had made Scripture to assert.*[7]

Burnet's grand theory did not fare well among natural philosophers, but it did spawn numerous alternative theories.

Notable among these was John Woodward's influential *Essay Toward a Natural History of the Earth*, published in 1695. In contrast to the saintly Steno, Woodward was by all accounts a self-promoting prima donna. Widely despised, but a genius in his own opinion, he was paranoid and uncharitable toward competitors and dismissive and unforgiving of critics. Famously vain, he reportedly had mirrors placed throughout his house so as to maximize opportunities to gaze upon himself.

Born in a Derbyshire village, Woodward apprenticed to a London linen draper. The king's physician noticed him there and virtually adopted the bright young man, eventually supporting his education and medical training. After receiving a doctorate from Cambridge, Woodward was appointed professor of medicine at London's Gresham College at the age of twenty-seven.

Woodward made his mark in natural history after he chanced upon shellfish entombed in solid rock in a Gloucestershire field. How sea creatures came to be encased in rock mystified him. Vowing to pursue an answer to the remotest parts of the kingdom, Woodward visited quarries and mines across Britain, noting anything memorable he came across and amassing a tremendous fossil collection. He sent off letters to natural philosophers inquiring about whether strata around the world contained fossils right up to the highest peaks. The same year he was appointed professor of medicine, he was elected to the Royal Society on the strength of his growing reputation as a fossil expert. So far, Woodward was building an impressive career.

The following year, in 1695, he published his essay, arguing that the Flood dissolved Earth's primitive crust, leaving no trace of the original world. Adopting Steno's principles, Woodward's ideas and the evidence he offered to back them up came from studying Britain's rocks and fossils. Convinced that fossils were the remains of organisms that perished in the Flood, he was more concerned with what the event accomplished than in how it came about.

Woodward was one of many natural historians whose homeland's landforms and geological features figured prominently in their thinking. It's no chance happening that English savants greatly influenced the explanation of fossil life. Their country, and much of its well-exposed coastline, is rich with fossils. I have no doubt that my own geological perspective on big floods would be quite different had I only stayed within several hundred miles of where I grew up in northern California and had never seen wonders like the Tsangpo Gorge and the Grand Canyon.

A good scientist also draws on the experience and observations of others, and despite his famous arrogance, Woodward borrowed Steno's idea that all strata were deposited as great horizontal sheets. He, too, argued that one could read the history of deformation from the orientation of formerly flat-lying rock. Like Steno, Woodward thought that topography formed during the same event that disrupted the rocks. Convinced that the only true philosophy was based on careful observation, he believed that his account of earth history confirmed that a great flood reshaped the world.

In Woodward's day, many natural philosophers accepted the idea that a mighty flood burst forth from a subterranean abyss. In keeping with then conventional wisdom, Woodward invoked a violent torrent to rip up and dissolve the planet's entire crust, mix it up, and suspend it in the raging waters. As the Flood receded, dense stuff settled out first, followed by lighter stuff. This resurfacing created the modern world, leaving fossils set in the resolidified detritus after the show was over.

To Woodward the problem was what triggered wholesale dissolution of Earth's surface. Inverting Newton's recognition that gravity held solid bodies together, he proposed that a temporary suspension of gravity dissolved the world into a chaotic mass. If God just flicked gravity off and then on again, it would create an instant deluge. Things settled out when gravity turned back on, sorted by weight into distinct layers—like those seen in rocks. Organic fibers, the very fabric of nature, would hold plant and animal tissue together, allowing fossils to remain intact in the resolidified earth. Then, after the Flood, some of the new layers settled and others rose, forming modern topography.

Woodward also appreciated the theological implications of a remodeled world. Foremost to him was how it revealed the second half of God's plan: "'Tis very plain that the Deluge was not sent only as an Executioner to Mankind: but that its prime Errand was to Reform and New-mold the Earth."[8] Before the Flood, the world was incredibly fertile, a perfect Eden where one need not plow or even plant to reap nature's bounty. But with idle hands having led to humanity's downfall, it made sense that God would remake the world into a place of no free rides, where eking out an existence required constant labor. Destroying the world, and mankind along with it, was the ultimate act of kindness.

> *For the Destruction of the Earth was not only an Act of the profoundest Wisdom and Forecast, but the most monumental Proof, that could ever possibly have been, of Goodness, Compassion, and Tenderness, in the Author of our Being.*[9]

For naturalists, Woodward's theory improved upon Burnet's in that it explained how fossils came to be incorporated into rocks. Still, Woodward caught even more flack than Burnet because he made a simple testable prediction—what we today consider a hallmark of good science. If Woodward was right, then the rocks and fossils within them would be ordered from densest on the bottom to lightest on top, reflecting the order in which things settled out.

Critics quickly pointed out how the heaviest fossils were often found on the surface rather than deep underground. Some objected to Woodward's idea of a turbulent globe-dissolving flood when the sedimentary strata it supposedly deposited showed signs of having settled down through tranquil water.

Woodward was considered brilliant by some, but his arrogance and habit of making enemies contributed to his undoing. In 1697, London physician John Arbuthnot gleefully skewered him in *An Examination of Dr. Woodward's Account of the Deluge*. It not only laid out problems with Woodward's theory but showed that the great blowhard had plagiarized Steno. Arbuthnot paired sections of Steno's obscure book with virtually identical sections from Woodward's popular essay. In passage after passage, Woodward had cribbed Steno without acknowledging his source. As it turned out, exposure of this act of intellectual theft helped promote Steno's ideas.

Arbuthnot's devastating critique stamped Woodward's account of the biblical flood as contrary to the laws of nature. How could the Flood have been violent enough to churn up and dissolve the entire surface of the world, and yet preserve both marine life and delicate plant fossils? Besides, Woodward's assertion that rocks and fossils were arranged on the basis of specific gravity was wrong. Arbuthnot himself had descended into a two-hundred-foot-deep pit in Amsterdam and found the density of the layers to be variable and not ordered by depth. Contrary to Woodward's theory, heavy layers lay on top of lighter ones. Fellows of the Royal Society of London corroborated Arbuthnot's findings, reporting that it was common to find denser strata overlying lower-density rocks.

Arbuthnot even conducted laboratory tests to disprove Woodward's basic contention, finding that when an oyster shell and an equal weight of metal powder were dropped into a tank of water, the oyster shell sank to the bottom first. His simple experiment showed that size and shape influenced how fast things settled. Arbuthnot calculated that Woodward needed a flood 450 miles deep to turn the world into a slurry of half earth and half water, a scenario he ridiculed with dry wit: "The Doctor should have calculated the Proportions of his Drugs before he mix'd them."[10] Just as with Burnet, Woodward's critics eventually took his theory down. That the rocks did not back up his story earned Woodward the distinction of having proposed one of the first grand geological theories to be formally refuted.

There was no shortage of subsequent fantasylike theories of the Flood, including one from astronomer Edmund Halley involving his namesake Halley's comet. When his predicted return of a comet to

European skies came true in September 1682, the popularity of comets surged among both the general public and natural philosophers. Two years later, Halley read a pair of papers to the Royal Society in which he argued that in dictating Genesis to Moses, God left out most of earth history. Fossils found far above the sea convinced Halley, like many before him, that the biblical flood was indeed global. Noting that God used natural means to carry out His will, and that forty days and nights of rain could not possibly submerge the highest mountains, Halley proposed that the shock from a comet passing close by Earth knocked the world off its axis, sending the oceans sloshing back and forth across the continents. The resulting devastation heaved the seafloor up into great piles, forming mountains and carving the topography we know today.

Even if the forty inches of rain that typically fell in a year in England's wettest counties instead fell each day for forty days and nights, it would only inundate coastal lowlands. So Halley drummed up another source in an act of God. A great vapor canopy God had originally placed above the firmament to enshroud the primordial Earth collapsed and dropped enough water to account for Noah's Flood. Three centuries later the founders of modern creationism resurrected this highly imaginative idea as their own vapor canopy theory.

Halley's second paper presented far more radical ideas. Maybe the comet hit more than four thousand years ago. Maybe such global calamities occurred many times in the past, and might even recur in the future. Periodic catastrophes might even be necessary to refresh Earth's surface once soils eroded and could no longer support life. He admitted to struggling with the theological implications of a world designed to require periodic destruction, and was terrified of what the church might think of his views. Less brave than Galileo, Halley refused to publish his papers and instead deposited them in the Royal Society's archives, with the proviso that they be published after his death.

Two years after Halley's address, in 1696, one of those in attendance, William Whiston, a Newton protégé and chaplain to the Bishop of Norwich, borrowed Halley's comet for *A New Theory of the Earth*. A combination of Newtonian physics, biblical interpretation, and occasional facts, Whiston's book also described the planet being knocked off its axis as it passed through the tail of a great comet. Whiston spun another tale from that point. Torrential rain from the comet's atmosphere opened the floodgates of heaven. The gravitational attraction of this near miss created enormous stresses as the rocky crust stretched and contracted under the influence of subterranean tides. As the crust cracked, the combination of torrential rain and water liberated from below scoured the world's surface. Then the floodwaters neatly drained back down into the abyss, leaving the churned-up mess to settle back into place much as Woodward had described.

Not everybody was impressed with such theories. Oxford astronomy professor John Keill published a critique of Burnet's and Whiston's arguments that condemned both men as "makers of imaginary worlds and loosers of imaginary floods."[11] Keill derisively labeled Burnet's book a "*Philosophical Romance*" because an originally smooth world bathed in perpetual sunlight would be uninhabitable.[12]

Rivers would not run on Burnet's perfectly smooth Earth. With no slope to drive the current, rivers could not flow. They would "stagnate and stink," making for "uncomfortable living."[13] With no rain and no flowing surface water, Keill thought that the land between the foul rivers would have been more like Hades than Paradise.

And Burnet's rocky crust could never float like clay flakes on an ocean of water. It would sink as soon as it consolidated. Besides, Keill noted, Genesis revealed that antediluvian society had iron tools, and thus Earth's original crust must have contained iron. Yet if Burnet was right, dense iron particles would have settled rapidly down through the abyss and would never have become incorporated into the crust in the first place. Keill dismissed Burnet as a victim of excessive imagination who used clever rhetoric to charm logic to sleep.

Still, that wasn't the biggest flaw in Burnet's theory. Had the warmth of the Sun been able to penetrate Earth's surface and heat the inner sea enough to crack the crust, it would have baked the planet's surface, raising insurmountable questions about Noah's Flood.

> *Certainly there could be no necessity for a Deluge in that case, except it were to cool the Earth again after such an excessive heat, which must have destroyed all the Animals, Plants, and Trees which were upon the earth, and have turned them into Glass.*[14]

Keill likewise demolished Whiston's theory by showing that there would not be enough pressure in a comet's tail to generate torrential rains. Keill further calculated that the gravitational pull of a passing comet would not deform a subterranean abyss, thereby burying yet another idea attempting to explain Noah's Flood.

Curiously, Keill the astronomer was a deeply religious natural philosopher not inclined to rationally explain the miraculous. He was comfortable with the Flood's being an event not amenable to scientific explanation. While the astronomer Keill preferred to invoke miracles to explain earth history, the cleric Burnet sought to demonstrate that it happened through natural processes.

Today, long after such fundamental ironies have been forgotten, seventeenth-century ideas still frame the essential arguments that creationists offer to reconcile geological evidence with their presumed reality of a global deluge. The key difference, of course, is that seventeenth-century philosophers did not blindly trust particular literal interpretations of scripture. They had faith reason would lead to enlightened interpretation of God's creation, as read from the pages of the book of nature—the rocks themselves.

As natural philosophers began to better understand the universe and its workings, attitudes toward mountains underwent radical change. Long seen as ugly, inconvenient, and dangerous, the Alps became Europe's prime tourist attraction by the end of the eighteenth century. At the same time, theologians gradually came to see mountains as beautiful natural cathedrals—spiritually uplifting examples of the magnificence of creation rather than evidence of a ruined world, the broken remnants of a wrecked paradise.

Geologists today tend to forget that the foundation of modern geology, Steno's deceptively simple idea that younger rocks lay on top of older ones, was introduced to help explain how Noah's Flood shaped the Italian landscape. Yet Steno's story remains one of the best examples of the complex interplay between geology and theology, setting off and setting up debates that continue to this day. Although Steno's greatest insight was that the present arrangement of the layers that make up our world can be used to read its history, his greatest impact was on shaping the views of generations of students he never met. The more natural philosophers applied Steno's rules to the geologic record, the more they discovered about how the rocks revealed a much longer story than the traditional biblically inspired history of the world.

5

A Mammoth Problem

TODAY, GEOLOGISTS KNOW THAT more than 99 percent of all animal species that have ever lived are extinct. You don't have to know any geology to know that trilobites, dinosaurs, and saber-toothed tigers no longer live among us (unless you count birds as modern dinosaurs). Given this, it makes no sense to argue that Noah's Flood explains the world's fossils. If that were the case, it would mean the Flood not only caused extinctions but killed off almost all the world's then living species—the very thing that Noah supposedly built his ark to prevent in the first place.

But in the opening days of the eighteenth century, naturalists and theologians alike were confident that extinctions had no place in God's plan. Almost everyone assumed that living examples of fossils would eventually turn up as more of the world was explored. Vigorous arguments continued to rage over how God triggered Noah's Flood, but after Steno, Burnet, and Woodward, natural philosophers increasingly interpreted internment of once-living creatures in rocks as compelling evidence of a divine disaster. After all, there was no way to know how old fossils were, no way to date when they had lived—or had died. Wasn't the simplest answer that they had died all at once?

If the only idea you have to explain rocks and topography is a big flood, then you will naturally tend to interpret the evidence you find in terms of a big flood for as long as you can. Even scientists today are not immune to interpreting evidence, at least initially, through the lens of prevailing ideas and their preconceived notions. Centuries ago, when natural philosophers learned of fossils near the crest of the Andes, they concluded that the biblical flood parked the bones of sea creatures within South America's highest mountains.

A problematic detail, however, muddied the waters—some fossils did not correspond to any known living species. One of the most striking fossils common in the layered (sedimentary) rocks of England were ammonites, snail-like marine animals with spiral shells characterized by distinctively crenulated partitions that created internal chambers. There was a dizzying array of different species and types of fossil ammonites, ranging in size from inches to several feet across. They were found throughout certain rock formations across southern England and were literally falling out of the cliffs to litter beaches along the English Channel. Yet nothing like them had ever been found alive anywhere in the world. Their closest living relative seemed to be pearly nautilus, an exotic chambered shell with simpler, noncrenulated partitions from the East Indies. Most natural philosophers shrugged off this problem, confident that someday someone would dredge a living ammonite up from the sea. They thought that only a flood of awesome power, the biblical flood, could have entombed on land creatures thought to live in the very deepest part of the ocean.

The views of diluvialists—those who invoked Noah's Flood to explain what they found in the rocks—dominated geological thinking until natural philosophers demonstrated that fossils were extinct and that Earth had a much longer and more complicated history.

A leading voice of the diluvialists was Johann Scheuchzer, one of continental Europe's great fossil enthusiasts. After completing a doctorate in medicine at Utrecht in 1694, he returned home to Zurich, where he eventually became a professor of mathematics. Insatiably curious about the natural world, Scheuchzer served as the secretary of a weekly club that held lively discussions on controversial topics such as whether the devil could physically seduce a woman and whether mountains were created along with the world or formed during Noah's Flood.

Scheuchzer's passionate interest in Swiss natural history led to extensive walking tours through the Alps. Accompanied by his students, he made geological observations and was the first to measure—by carrying a barometer up a mountainside—how air pressure changed with altitude. Fossils especially fascinated him. He had been taught they were mineral oddities whose origin could be explained by physics and chemistry.

When Scheuchzer read Woodward's essay, he realized that fossils really were ancient creatures. Right under his nose, entombed in his own rock collection, were the remains of snails, seashells, fishes, and plants. This revelation prompted his own landmark work in 1708, *The Fishes' Complaint and Vindication*, in which Scheuchzer lampooned the still popular idea that fossils were inorganic objects that just happened to resemble real creatures. He shaped his narrative from the point of view of a fossil fish who complained in formal Latin about not being recognized as an innocent victim of the flood sent to destroy mankind.

> *"We, the swimmers, voiceless though we are, herewith lay our claim before the throne of Truth. We would reclaim what is rightly ours. . . . Our claim is for the glory springing from the death of our ancestors . . . carried on the waves before the Flood. . . . We bear irrefutable witness to the universal inundation."*[1]

Scheuchzer's fossil narrator righteously demanded the dignity of being recognized as having suffered alongside mankind during the Flood. Speaking for innocent marine creatures that died when receding floodwaters stranded them on dry land, it added insult to injury to deny that their own bones testified to their existence. The fossilized spokesman introduced detailed illustrations of marine fossils that any fisherman would recognize as the remains of familiar animals.

The year after his fossil fishes spoke up, Scheuchzer published *Herbarium of the Deluge*, a collection of botanical prints illustrating plant life purportedly fossilized as a result of the Flood. This collection of striking images showed exotic plants set in stone, offering a window into a world before our own. That ferns and tropical plants had been growing in Europe drew open the curtain of time to reveal a radically different world.

Seduced by what he saw as fossilized postcards of life before the Flood, Scheuchzer kept looking for more flood victims. The limestone quarry at Oenigen, in the Alps near the west end of Lake Constance, gave him access to fossil fish, bullfrogs, snakes, and even turtles. He saw these fossils, now known to date from the Miocene epoch (ten to twenty million years ago), as relics of Noah's Flood deposited along with the rest of the world's sedimentary rocks. Then in 1725 stone workers at the quarry unearthed part of an unusually large skeleton and shipped it off to Scheuchzer, who promptly interpreted it as another flood victim. What better testimony to the veracity of the biblical flood than the bones of a drowned sinner?

Naming this unlucky fellow *Homo diluvii testis* (man who testifies to the Flood), Scheuchzer sent off descriptions of his incredible find to British, French, and German journals and published a short book that shared the fossil's name. Scheuchzer's discovery of a human witness to the Flood not only showed that a world of sinners drowned but that they were giants, just like the Bible implied when it said "there were giants in the earth in those days" (Genesis 6:4). Scheuchzer had a ready answer for the dearth of human remains in the rocks laid down by the Flood. The bones of innocent animals were to remind us of their sacrifice, whereas the rarity of human remains confirmed that sinners deserved condemnation to eternal oblivion.

Convinced he had found proof of Noah's Flood, Scheuchzer spent his last years compiling his *Sacred Physics*, in which he sought to harmonize natural history with scriptural truths. He proposed that the fountains of the deep had burst forth when the hand of God literally reached out and applied the brakes to Earth's rotation, stopping the world dead in its tracks, splitting continents apart and spilling out subterranean seas to produce the biblical flood.

That idea didn't catch on, but Scheuchzer's human flood victim was a sensation. A museum in Haarlem acquired *Homo diluvii* to show it off to the faithful. Although natural philosophers decided within a few decades that it probably was just a big fish, it remained a popular attraction until 1812, when the prominent French anatomist Georges Cuvier, whom we'll meet shortly, authoritatively declared it otherwise. Ironically for a talented naturalist, Scheuchzer's faith that

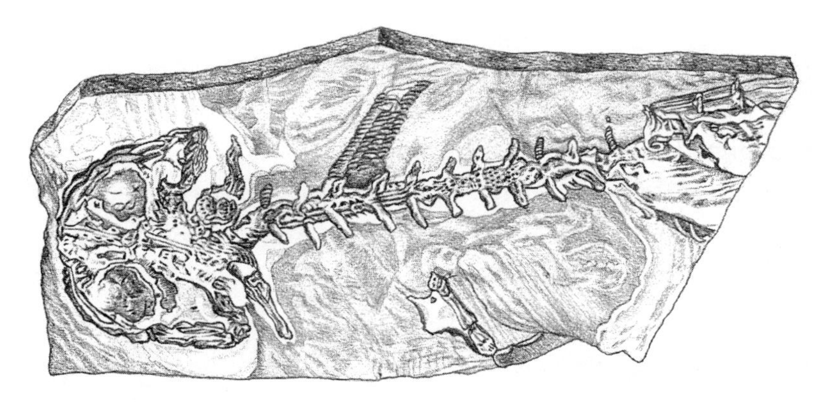

Homo diluvii, *the fossil Johann Scheuchzer interpreted as a victim of Noah's Flood (by Alan Witschonke based on plate XLIX of Scheuchzer's* Sacred Physics *(1731)).*

the geologic record told the story of Noah's Flood led him to the colossal blunder he is still lampooned for today. As Cuvier pointed out, Scheuchzer's flood victim was a giant amphibian.

These were not the only strange bones attributed to Noah's Flood. All across Europe large fossils were publicly displayed as the remains of the giants mentioned in scripture. Scheuchzer didn't get it all wrong, because he pointed out that the enormous stone teeth of those purported to have drowned in the Flood actually belonged to something more like an elephant than a person.

The expansion of European power and influence in the eighteenth century led to the discovery of giant bones in Siberia and North America. In 1692, Peter the Great's envoy to China, Ysbrand Ides, found frozen tusks and hairy elephant carcasses exposed in a Siberian riverbank. His report claimed these behemoths lived before the biblical flood, their frozen hulks preserved by a frigid post-Flood climate.

Further expeditions returned to St. Petersburg with the partial remains of huge creatures that the indigenous Siberians called "mammut," a name European tongues promptly changed to mammoth. Within a few decades, such discoveries convinced natural historians that there was an abundance of fossil elephants in Siberia, a place too cold for African animals to survive today. With the closest living elephants located in India, natural historians tended to interpret the Siberian bones as those of creatures swept north from Asia by a great flood, in much the way remains of African elephants were thought to have made it to Europe.

Similar finds in North America were also attributed to Noah's Flood. Large bones found along the banks of the Hudson River in upstate New York were thought to be those of an antediluvian giant. Discovered eroding from a hillside in 1705 near Albany, a six-inch-tall, two-and-a-quarter-pound tooth and a seventeen-foot-long thighbone convinced Cotton Mather, of Salem witch trial fame, that giants really did drown in the Flood. Dug out from the base of the hill, the great thighbone crumbled away when exposed to the air. Mather was convinced that the more durable four-pronged tooth looked like a human molar, only much bigger. All who saw it thought that this was a victim of Noah's Flood. Based on the size of such bones, one authority estimated that Adam was well over a hundred feet tall. Mather's giant bone, however, was probably a mammoth bone.

Mather was enthralled with his fossil finds, and in November 1712, he wrote the first of a series of letters to the Royal Society in London to bring to the attention of scholars these New World curiosities from the time of the Flood. He also reported accounts of giant bones discovered in South America, convinced that they, too, were proof of Noah's Flood.

> *Below the Strata of Earth, which the Flood left on the Surface of it, in the other Hemisphere, such Enormous Bones have been found, as all Skill in Anatomy, must pronounce to belong unto Humane Bodies, and could belong to none but GIANTS. . . . The Giants that once Groaned under the waters, are now found under the Earth, and their Dead Bones are Lively Proofs of the Mosaic History.*[2]

The flood that buried giants appeared universal in Mather's mind and fit in well with his belief that Moses described Noah's Flood as a global event. In 1721, Mather wrote *The Christian Philosopher*, the first systematic book on science published in America. Invoking fossils as direct evidence of a global flood, it was dedicated to the argument that reason supported faith.

Not everyone was convinced that giant bones were the bones of giants. Around 1725, English botanist Mark Catesby visited Stono, a large plantation near Charleston, South Carolina, to examine gigantic teeth that slaves had unearthed from a swamp. While the plantation owners thought that the colossal molars were the remains of a giant that drowned in the Flood, the native Africans who had found them swore that they were dead ringers for elephant teeth. Catesby scandalously shocked his hosts by agreeing with their slaves. Unlike the plantation owners, he had seen elephant teeth on display in London.

Catesby got closer than Mather to deducing the true origin of giant bones, but it was not until the next decade that the bones were pegged to mammoths. Exploration of the vast, unexplored wilderness west of the Appalachians proved to be key. In 1739, a French military expedition traveling from Niagara to the Ohio River discovered enormous bones at a salt lick near the river. Recognizing the value of this cache of fossils, the commander of the expedition sent a fossil tusk, a giant femur (thighbone), and several huge teeth down the Mississippi and on to Paris. The site became famous as Big Bone Lick. While some natural philosophers believed that these mammoths were a new species larger than modern elephants, others thought that the difference in size between modern and fossil bones was no greater than the degree of variation in size among modern elephants.

Thomas Jefferson, for one, was convinced that North American mammoths were the same species as Siberian mammoths, distant hairy cousins of tropical elephants adapted to life in cold climates. Jefferson had an intense interest in the natural world and plants and animals. The prospect of living behemoths among the fauna of the American wilderness thrilled him. As governor of Virginia, he was familiar with the fossil discoveries at Big Bone Lick, which then lay within the expanded borders of his state. In 1781 Jefferson published his *Notes on the State of Virginia*, the only full-length book he wrote, in which he described mammoths as larger than an elephant and told how Native Americans considered the giant bones at Big Bone Lick to be those of the "Big Buffalo," the largest of animals. He related native stories that told of how the giant teeth from Big Bone Lick belonged to an enormous carnivore that still roamed America's unexplored northern wilderness.

Jefferson collected examples of the richness, vigor, and brute size of American wildlife, even displaying a taxidermied bear inside the White House, to counter French claims that European animals were superior to American fauna. What could better make the case for the superiority of American animals than an elephant-sized predator? It would be a powerful symbol of his new nation, embodying the independence and strength of the American character. Trappers and explorers were still finding new, exotic animals west of the Appalachians. Might not someone find a living mammoth? This was the same argument European savants used to rationalize why no one had ever seen a living ammonite. But, unlike ammonites, mammoths could not be hiding in the deep sea.

Across the pond, scholars were starting to doubt that mammoths were still alive and well. Near the close of the century, in 1796, Georges Cuvier, a professor of natural history at the College de France and the Muséum National d'Histoire Naturelle, compared bones of mammoth carcasses to those of African and Indian elephants. Mammoths matched neither living species. But if Noah saved all the animals, how could these fossils represent extinct animals? Was this possibly evidence of animals that lived and died long before the Flood and that inhabited a world much older than the one laid out in Genesis?

A lifelong churchgoer, Cuvier was born into a Lutheran family in the French-speaking German Duchy of Württemburg. By the time of the French Revolution, he had built a reputation as an expert in animal anatomy, studying marine organisms while working as a tutor for a family of nobles in Normandy. When France annexed his hometown, he moved to Paris, where he was appointed understudy to an aging professor. He had a unique talent for understanding the relation between invertebrate form and function, and rapidly rose to prominence in scientific circles. Cuvier also served as the vice president of the Bible Society of Paris. At the natural history museum, he had the opportunity to see collections of fossils from all over the world.

When the revolutionary armies of France swept through what is now Belgium, an official team of trained specialists, including a naturalist, accompanied them to plunder useful or valuable objects. Most of the team focused on acquiring the best crop varieties and agricultural machinery. The naturalist had an eye for extraordinary fossils and returned to Paris with loot fit for a king.

As a hundred and fifty crates of specimens from France's new conquests to the east arrived at the museum in Paris, so, too, did Cuvier. It was to be a turning point in his thinking and career. He found two elephant skulls among the samples that were unpacked in the auditorium, one from southern Africa and the other from Ceylon (Sri Lanka), off the southern coast of India. Cuvier carefully measured and analyzed these skulls alongside those of Siberian mammoths and found that they were from distinctly different species. The conclusion was clear—the mammoth skulls resembled no living species.

He also compared the teeth of elephants and mammoths with those from Big Bone Lick. The grinding surface of the teeth of one of the American specimens was covered with unusual knobs that resembled small breasts. This was a different species than the Siberian mammoths, which had raised ridges on their teeth. He named the peculiar American specimen "mastodonte," breast-tooth.[3]

Cuvier concluded that there were three kinds of elephants. There were the modern African and Asian species, the Siberian mammoth (which also had lived in North America), and the mastodon, which was only found in North America. Although they were all herbivores, mammoths ate grass and mastodons ate woody shrubs and trees. The uncomfortable fact that both were extinct opened the door to seeing plants and animals as organic beings subject to change.

How many other species were extinct? When did they die off, and what was the world like when they lived? Cuvier put his expertise in comparative anatomy to work by analyzing fossils to reconstruct the inhabitants of vanished worlds. He found that whole faunas preserved in stone were distinct from living species. His findings convinced him that ancient worlds were radically different from the one he knew. The world had a complicated and dynamic history. Species came and went through time.

Cuvier thought that the story told by rocks and fossils roughly paralleled a nonliteral reading of Genesis. He also thought that the story of Noah's Flood was the story of some type of recent global catastrophe, which had wiped out large mammals known only through their fossils, like mammoths. Cuvier maintained that the legends of the ancient Egyptians, Greeks, and Jews all pointed to a grand disaster immediately prior to the dawn of human history.

Cuvier sought to marshal observable facts to trace the history of the world and to understand the sequence of grand disturbances, or revolutions that had punctuated earth history. It seemed that life had turned over every now and then throughout geologic time. The story Cuvier read was one that began with initial life-forms and transformed into a world of ammonites and sea life. Then, a whole succession of worlds with novel terrestrial faunas arrived, with people arriving in the most recent, modern world.

Offered the chance to accompany Napoleon's expedition to Egypt, Cuvier chose to stay close to the collection of the museum. He preferred to have specimens come to him and issued an appeal for collectors to send fossils, drawings, or descriptions for him to assess. In return, he offered to authoritatively identify the bones, a skill that few others in his day possessed. Cuvier's masterful ability to relate the structure of organisms to their biological function netted him a role as a scientific referee on issues related to vertebrate anatomy. Today, he is known as the founder of vertebrate paleontology.

In the first public summary of his research, Cuvier treated fossils as if they were all the same age. The bones of fossil elephants (mammoths) were evidence of a previous world destroyed by some kind of catastrophe. Later, as he came to realize that different geological formations held distinctive fossils, he recognized that the fossils in the older beds were progressively different from the modern fauna.

As he continued to amass specimens, Cuvier increasingly recognized patterns in the organization of life through time. Ammonites were found exclusively in the lower and therefore older formations, mammoths were found in the highest and most recent formations of surficial debris. Human bones were not found as fossils. If fossils truly represented extinct plants and animals, and not just species hiding out in the deep sea or in unexplored wilderness, then Earth had a distinct history in which life approached the form of the present fauna through the turnover of species unlike any known today. Cuvier's skills, intellect, and intuition combined to lead the way forward in piecing together earth history. His advances rivaled those of any other scholar up until that time.

Cuvier speculated that extinctions happened during violent geological revolutions, sudden disasters for which he invoked the well-preserved bodies of mammoths as evidence: "In the northern regions it has left the carcases of some large quadrupeds which the ice had arrested, and which are preserved even to the present day with their skin, their hair, and their flesh."[4] In Cuvier's view, developed from the great number of fossils he studied, a not quite six-thousand-year-old Earth was simply inadequate to accommodate the diversity of fossil life. Certainly, one great flood was not enough to explain earth history. "Life, therefore, has been often disturbed on this earth by terrible events—calamities which, at their commencement, have perhaps moved and overturned to a great depth the entire outer crust of the globe."[5]

We now know of at least five mass extinctions in the geological past, and biologists say another one is under way as we wipe species off the planet 100 to 1,000 times faster than nature did before we started helping out. Since the evolution of life on land, several events have killed off over half of all animal species. Every school kid learns that dinosaurs died off and mammals began rising 65 million years ago during the great Cretaceous-Tertiary extinction event. The less well-known, but far deadlier, Permian-Triassic extinction event 251 million years ago killed off almost all of the animal species on Earth, ending the age of trilobites and setting up the rise of dinosaurs. More recently, the last glaciation of the Quaternary Period (the so-called ice age of the past several million years) saw the demise of megafauna, like mammoths, and ushered in a modern world increasingly dominated by people. When viewed through the geologic record millions of years from now, the modern extinction event we are living through may well look similar to past grand catastrophes that ended ancient worlds.

After Cuvier, the drive to find evidence for Noah's Flood in the rocks was well and truly dead, although modern creationists would later resurrect the idea. While natural philosophers were long wedded to the idea that fossils confirmed the biblical account of a great flood, once they established the reality of extinctions in the geologic record, it showed that Noah's Flood could not have deposited all the world's fossils. They then shifted to looking for the signature of the Flood in the overlying unconsolidated deposits of gravel and boulders. This new view helped natural philosophers and theologians alike accept a pivotal reinterpretation of the Bible, one that made room for a new concept of time—time enough that fossils need not have all died, or lived, at the same time. Thanks to a Scottish farmer, today we know this idea as geologic time.

6

The Test of Time

THIRTY MILES EAST OF Edinburgh lies Siccar Point, a holy site of sorts. The farmer whose fields surround it is said to complain about an endless stream of geologists trampling his turnips. Rock hounds plague this windswept headland because it's celebrated as the place where Scottish farmer James Hutton discovered geologic time—the place he found the key to unlocking time enough for geological forces to reshape the world. Tucked in along the rocky shore below the turnips are the clear signs of two rounds of mountain building, erosion, and deposition recorded in two sandstones, one gray and the other red.

On a rare sunny Scottish day six of us pulled up at the trailhead and parked just out of view from the farm. We skirted the fields and walked toward the sea cliff, passing by the ruins of a crumbling building amid glowing yellow gorse bushes. I could see striking beds of red sandstone diving down toward the sea to the west. To the east lay planed-off vertical beds of gray sandstone exposed along the shore. Walking out to the headland, we stood above where the

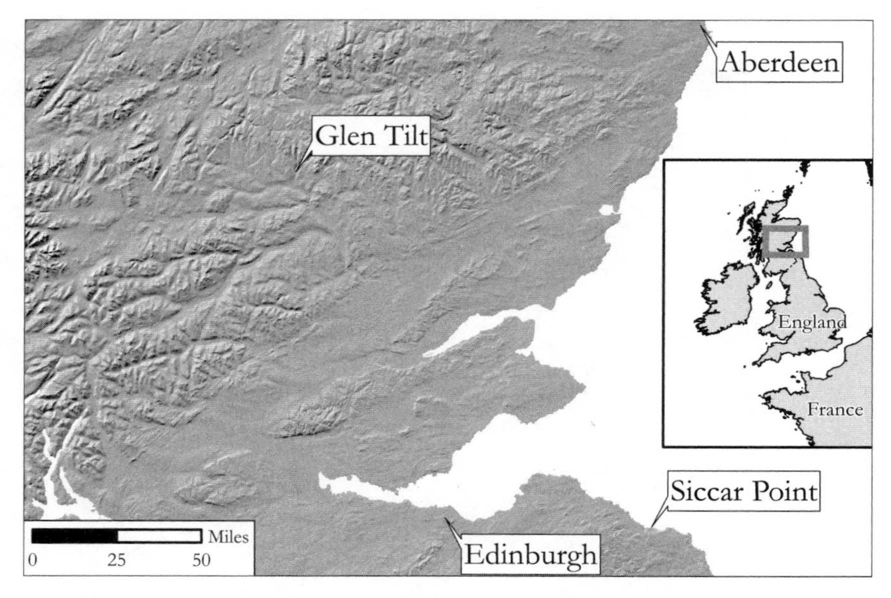

Map of Siccar Point, Scotland, showing its position on the coast east of Edinburgh.

two rock formations should meet before starting down a steep grass-covered slope pitching off to the surf below.

At the bottom lay a jewel of an outcrop. The two rock formations sat there just as textbooks showed. Here, in front of me, were the rocks that helped inspire geology's core concept of deep time, that the world is billions of years old. Over lunch I read the story in the rocks, laid out plain as day.

The older gray sandstone formed as debris eroded off an ancient upland and settled to the bed of an adjacent sea until the sand eventually lay buried deep enough that heat and pressure turned it into solid rock. Then, something caused the rocks to buckle, lifting them back above sea level and tipping them into their now vertical orientation. Gazing along the shore, I could see how the contact between the two sandstones defined the surface of an ancient valley carved into the gray sandstone. As this new land sank back down beneath the waves of an ancient sea, red sand settled on top, eventually accu-

Hutton's unconformity at Siccar Point showing the inclined beds of the Silurian Old Red Sandstone truncating vertical beds of Devonian graywacke sandstone (by Alan Witschonke based on a photograph by the author).

mulating into enough of a pile to turn it, too, into bona fide rock. After all that, another round of tilting and uplift brought the works back to the surface, where waves peeled the cliff back to expose a low shelf of red sandstone dipping out to sea at a jaunty angle and truncating the underlying vertical beds of gray sandstone.

When Hutton discovered this outcrop in 1788, it confirmed his suspicion that mountains could be recycled into sand and remade into new rock. I had the advantage of having my colleagues from the University of Edinburgh explain how the gray rock, four-to-eight-inch-thick beds of sandstone separated by thin layers of mudstone, recorded erosion of the mountains that formed the geologic suture from the closing of the ancestral Atlantic Ocean. This collision united England and Scotland 425 million years ago during the Silurian Period, several hundred million years before the days of the dinosaurs. The upper formation, the Old Red Sandstone, formed when the younger Caledonian mountains eroded 345 million years ago in the Devonian Period, with the resulting sand deposited in what is now modern Scotland. The other half of the sandstone derived from erosion of the Caledonian mountains lies across the Atlantic, in New England, as the Catskill Formation in New York and Maine. The present far-flung distribution of the two halves of the red sandstone records the reopening of the Atlantic Ocean well after the life and death of the mountains testified to by the rocks themselves.

Although I'm well versed in thinking about geologic time, I still have a hard time grasping how long it must have taken to raise and erode a mountain range, deposit the resulting sand in the sea, fold up the seabed into another mountain range, and then erode it all back into a new ocean. Siccar Point stands as a natural monument to the unimaginable expanse of time required to account for geologic events.

Of course, in Hutton's day general consensus placed the world at a mere six thousand years old. The crazy notion of a world old enough to be shaped by the slow accumulation of day-to-day change was beyond radical, it was dangerously pagan.

Nowhere does the Bible say, "the earth is six thousand years old." This curious belief comes from literally adding up years gleaned from biblical chronology to arrive at how far back the world was created. The second-century historian Julius Africanus was the first Christian to date the Creation by drawing on Egyptian, Greek, and Persian histories. His urgency in dating the dawn of time stemmed from the belief that Christ would return to begin his thousand-year reign before the end of the world precisely six thousand years after it all began. The only way to be sure about when the world would end was to figure out when it started.

Adding up the ages of Adam's descendants listed in Genesis, Julius convinced himself that 2,261 years passed between the Creation and Noah's Flood. He then summed up the ages of Noah's descendants and used extrabiblical sources to determine the dates of key events such as when Moses led the Jews out of Egypt and the destruction of the Temple in Jerusalem. In this way, Julius determined that Jesus was born precisely 5,500 years after God created the world. Adopting the tradition attributed to the prophet Elijah that the world would only last a thousand years for each day in the week of Creation, Julius predicted that Christ would return to end the world in 500 AD. His *Chronologia* served as the model for subsequent biblical chronologies, both in approach and motivation.

Centuries later, medieval and Renaissance chronologists generally agreed with Julius that the world would last a thousand years for each day of Creation. They disagreed about when the countdown to the end started, repeatedly pushing the date by which the world would end further into the future as predicted apocalypses came and went without incident. By the end of the seventeenth century, there were more than a hundred biblical chronologies to choose from that set differing dates for the beginning and end of everything.

The most venerated biblical chronology is Bishop Ussher's influential *Annals of the Old Testament*. Published in 1650, it revealed Sunday, October 23, 4004 BC, as the date of Creation. Archbishop of Armagh and Primate of All Ireland, James Ussher was a confidant of Charles I, with an international reputation as a brilliant scholar and one of the largest personal libraries in western Europe. Ussher's prestige was such that he was buried with full honors in Westminster Abbey.

Ignoring Egyptian and Chinese histories that extended back before his preferred date for the Creation, Ussher concluded that Noah's Flood occurred 1,656 years after the dawn of time. Noah and company embarked on Sunday, December 7, 2349 BC, spent a little over a year aboard, and disembarked on December 18 the following year.

How did he establish the year of Creation from the Bible? Like Julius, Ussher tallied up the lifespans of the biblical patriarchs listed in the unbroken male lineage of who begat whom from Adam to King Solomon. To fill in the gap from Solomon to the birth of Jesus, he had to cross-reference biblical events with those of a known age from Babylonian, Persian, or Roman history. Ussher also had to choose which translation of the Bible to use, as the genealogy in the Greek Bible pushes the date of Creation back almost another thousand years. Finally, he corrected for the awkward problem that the first-century Roman-Jewish historian Josephus indicated that Herod died in 4 BC, and thus that Jesus could not have been born after that since the Bible says that Herod tried to kill the newborn Jesus.

How could Ussher pinpoint the day it all started? He used reason. God rested on the seventh day after the Creation, and the Jewish Sabbath is traditionally Saturday. So, counting backwards six days from Saturday, God started making the world on a Sunday. Assuming that the Creation began near the autumn equinox, Ussher probably used astronomical tables to determine that the equinox occurred on Tuesday, October 25, making Sunday, October 23 the best fit for the day it all began. However he came up with it, in 1701, the Stationers' Company inserted his 4004 BC date of Creation into a margin note for a new edition of the King James Bible. From then on, his calculated guess as to the age of the world became gospel for many Christians.

Despite the popularity of Ussher's chronology, dozens of biblical analysts offered competing claims. Their disagreements illustrate the inherent difficulty in pinning down the meaning of even literal interpretations of the Bible. Depending on the reader and what else he or she brings to the table, two people may arrive at different meanings. After Steno, natural philosophers began to pursue increasingly independent approaches, piecing together earth history directly from reading the rocks.

The influential Baron Georges-Louis Leclerc, Comte de Buffon and director of the botanical gardens in Paris, argued that the world was at least ten times older. Born into a family of wealthy French aristocrats, Buffon inherited the family fortune at a young age, giving him the freedom to study law before he turned to mathematics and natural history. When he became keeper of the king's garden in Paris in 1739, he converted it into a center to pursue his research interests.

In 1749, after a decade of study, Buffon proposed that Earth was created when a comet smashed into the Sun and knocked loose a molten fireball. The cooling of this piece of the Sun to form our world was described in the first installment of his massive thirty-four-volume *Histoire Naturelle.* After the flaming blob cooled into a rocky satellite, a universal ocean receded to expose the continents. Buffon denied that Noah's Flood ever occurred and suggested that animals evolved based on otherwise enigmatic vestigial organs that served no apparent purpose, like the sightless eyes of a mole and the wings of flightless birds.

Two years later, in January 1751, the theological faculty of the Sorbonne sent Buffon a letter calling him out for more than a dozen reprehensible ideas. Among Buffon's heretical notions were that currents scouring the bed of the primeval ocean shaped mountains and valleys, that topography was made by erosion rather than by God, and that eventually erosion would grind mountains down to sea level. Faced with the same choice that confronted Galileo, Buffon chose to recant and keep his prestigious position. He renounced everything in his book "respecting the formation of the earth, and in general all which may be contrary to the narrative of Moses."[1]

Shaken but undeterred, Buffon experimented with how long it took to cool spheres of molten metal. He determined that the first day of Creation had to have lasted more than twenty-five thousand years for the planet to cool to the point where water could settle on it. Based on rainfall rates, he calculated that the second day must have lasted ten thousand years to build up the primordial seas. His concluding estimate was that the world must be about 75,000 years old to have cooled to its present temperature. This time, when Buffon included this estimate in his *Introduction to the History of Minerals* in 1775, he escaped theological condemnation.

Three years later, Buffon expanded on the idea of an ancient Earth in his *Epochs of Nature*. He argued that the days of Creation were figurative and corresponded to geological ages, while cautiously refraining from publishing his own opinion that the world was millions of years old. The first of his great epochs saw the formation of Earth and other planets. During the second epoch Earth's rocky interior consolidated, releasing volatile substances to create the atmosphere. During the third epoch, about thirty-five thousand years after the planet formed, continent-covering seas deposited stratified rocks, coal, and marine fossils. Rushing currents circulating on the bottom of this great sea carved modern topography. Volcanoes became active in the fourth epoch. He offered Siberian fossil elephants (mammoths) as proof that even the poles enjoyed a tropical climate during the fifth epoch. In Buffon's sixth epoch the modern continents formed as the intervening land collapsed to form ocean basins. Finally, the arrival of mankind ushered in the world we know roughly six thousand years ago.

Although he did not grant Noah's Flood any place in his geologic history, Buffon did point out that there was no conflict between Genesis and geology if one did not take the days of Creation literally. He thought, just as some theologians had argued, that Genesis was written for uneducated people and should not be interpreted literally on matters pertaining to earth history. It was never intended to convey scientific truths.

Again, the church remained silent, torn by internal controversy over how to interpret Genesis. Unlike Galileo, this time Buffon escaped censure because influential theologians were themselves toying with the notion of an old Earth. Catholic opinion in France was divided about how to interpret Genesis. Even those in positions of authority were now willing to consider the idea that the six days of Creation might refer to geological ages.

Among Buffon's correspondents was Joseph Needham, the first

Roman Catholic priest elected to Britain's Royal Society. In embracing Buffon's view that each day in the week of Creation represented more than twenty-four hours, Needham pointed out that even sixty million years represents an infinitesimal portion of eternity. Theologians were starting to waver on a six-thousand-year-old Earth.

As the idea that geologic time involved more than a few thousand years became reasonable, Abraham Werner, a charismatic professor at the Freiberg Mining Academy, began popularizing the idea that the rocks revealed that earth history consisted of four periods. Werner's father, a Saxon foundry inspector, had passed on to his son a keen interest in minerals, and at the age of twenty-five Werner published an influential field guide that landed him a professorship at the Freiburg School of Mines. Five years later he offered the first course in historical geology. A gifted lecturer, Werner's influence grew as his students dutifully spread his ideas about geologic history across Europe.

A lab man who wanted to understand earth history from the study of minerals and rocks without all the bother of fieldwork, Werner adopted Buffon's view that our planet formed when a stray comet smashed into the Sun, spinning off a fireball that slowly cooled to become covered by a universal ocean. He proposed that the primary (crystalline) rocks precipitated from this global sea, accounting for marine fossils found high in mountains. Neptunists, as Werner's disciples were known, attributed deposition of the secondary (layered) rocks to material settling slowly to the bottom of the drying sea. They saw the signature of Noah's Flood in the sculpting of topography, and the deposition of the tertiary rocks that were made of gravel, sand, and clay derived from erosion and redeposition of the primary and secondary rocks. On top of all this was a fourth, or quaternary, level of unconsolidated sand and gravel eroded off uplands by running water, like the deposits of modern rivers. In short order, these four divisions were found to adequately describe the rocks of other mountain ranges, like the Apennines and Caucasus.

As this crude geological system began to formalize the basis for evaluating the thickness, lateral extent, and relative age of rock formations, it became apparent that irregular boundaries (unconformities) separated geological eras. And yet individual layers within the secondary rocks could be traced across Europe. Delicate layers just a few centimeters thick could be traced across tens of kilometers, something impossible to attribute to a chaotic deluge that ripped apart and mixed up the world's surface in the way that Burnet and Woodward had imagined. Werner's dominant influence on geological thinking meant that layered rocks were no longer all thought to date from the Flood. Now it was just the tertiary rocks and the form of the land itself that testified to the Flood.

A few years later, in 1788, James Hutton's startling discovery on a windswept stretch of Scottish coast went a step further in proving that earth history was more complicated than allowed by a literal reading of Genesis. At least two rounds of deposition and erosion were required to account for the deposition and deformation of the sandstone beds at Siccar Point—meaning that there were either two independent rounds of Creation, or Earth reshaped itself every now and again.

The son of a successful merchant, Hutton lived comfortably while studying at the University of Edinburgh. Upon graduation in 1743, at the age of seventeen, he apprenticed to a solicitor, offsetting the drudgery of copying wills and contracts by distracting coworkers with occasionally calamitous chemistry experiments. By the end of the summer Hutton's experiments had exhausted his employer's patience. That fall he reenrolled at the university, this time as a medical student. In 1747 he left Edinburgh to continue his studies, starting in Paris and finishing two years later with a medical degree from the University of Leiden (Steno's alma mater).

Despite his medical training, Hutton never seriously considered practicing medicine. Insatiably curious, he continued studying chemistry before turning to geology. Inspired by a favorite experiment, Hutton started a company with a former classmate to use chimney soot to make sal ammoniac (ammonium chloride). This key component of metalworking flux otherwise had to be imported from Egypt. The scheme was brilliant. Chimney sweeps were thrilled to get rid of soot, and metalworkers were glad to have an affordable and reliable supply of an essential ingredient. In combination with his inheritance, the profits meant Hutton need not work, which left him plenty of time to pursue his many other interests.

At first, Hutton devoted himself to his family's farm. Set on 140 acres just north of the English border, it lay on some of the best land in Scotland, where rolling hills carved out of volcanic rock produced rich, fertile soil. In contrast to Darwin's epic voyage around the world, Hutton began forming his radical ideas about the age of the world by watching the dirt wash off his fields.

As he learned to read the land, he translated his love of chemistry to agriculture, developing ways to use calcium carbonate to enhance soil fertility. He also tried to retain the soil eroding off his bare, plowed fields by enclosing them behind stone walls. Stacking blocks of sandstone quarried from nearby hills, Hutton couldn't help but recognize the similarity between the mineral grains leaving his fields and those that composed the rocks he piled.

There, in his hands, below his feet, and before his eyes, lay the keys to a grand cycle in which rocks eroded and the resulting sediment was deposited elsewhere and buried deep enough to reform into new rock. Most rocks in Britain are made of sediments eroded from somewhere else, and everywhere above sea level is eroding. Neither idea was new—Leonardo had long before recognized the nature of sedimentary rocks, and most farmers were familiar with erosion. But Hutton did something new: he put these ideas together, seeing them as two halves of a grand cycle. Here was the foundational insight behind his radically original concept of deep time.

Such a cycle presented a dilemma. Without a way to restore eroded material, the soil would eventually disappear and, along with it, the fertility of the land, something a benevolent creator would not allow. What could refresh the land after erosion wore it down?

After setting up his farm, Hutton moved back to Edinburgh in 1767. He arrived in a city on the cusp of an intellectual renaissance. The Scottish aristocracy that backed Bonnie Prince Charlie's failed attempt to claim the throne had been purged, dismantling class distinctions and ushering in a new egalitarian spirit that fostered innovative thought. The new intellectual culture that sprang from the ruins of Edinburgh society nurtured Hutton's curiosity and interests.

At the time, most natural philosophers thought rocks precipitated out of Werner's drying primeval ocean in a global version of those grow-your-own crystal sets. But Hutton's continual experimentation with mineral chemistry convinced him that rocks contained a lot of material that would not dissolve in water. How could rocks precipitate out of a drying sea if they could never be dissolved in the first place? And if Werner's conventional wisdom about how minerals formed was wrong, then what could be responsible for solidifying rocks? Hutton theorized that the combined effects of heat and pressure offered the only viable alternative. Both would be available at the bottom of a pile of sediment—as long as the pile was thick enough.

In 1784 the newly chartered Royal Society of Edinburgh invited Hutton, then nearly sixty, to present his theory of the Earth, forcing him to gather his thoughts into presentable form. He did not give his own lecture, whether due to illness or a bad case of nerves. His best friend, Joseph Black, who had recently discovered carbon dioxide, graciously read it—the tradition being that lectures were written up in advance and simply read aloud at the meeting. Black presented Hutton's ideas about layered rocks being made of sediment eroded off of previous land, and how heat and pressure were required to form rocks, as well as the case for rejecting Werner's ideas about rocks precipitating from an ancient sea. Ignoring the Bible and the Flood, Hutton had inferred that the world was unknowably old. Instead of a grand catastrophe to explain the world, he invoked the subtle day-to-day action of wind, rain, and waves that he himself had observed.

Four weeks later, Hutton personally read a second lecture. He finished his critique of Werner's theory and focused on how to get stratified layers of rock back to the surface after they solidified at the base of a thick pile of sediment. If rocks just precipitated from a shrinking ocean, then they should all lay horizontal. Yet it was well known that some layered rocks lay steeply inclined. Instead of invoking worldwide collapse during Noah's Flood to explain the tilted layers (as Steno had), Hutton literally turned the problem on its head and proposed a different action—Earth's internal heat and volcanic action was what deformed rocks. The key to his argument was how granite veins cut across layered rocks. If, as he thought, granite began as molten rock that rises up from the overheated base of a sedimentary pile, granite veins in cracks and fissures should cut across the layers in the rocks they pushed up through before cooling. Hutton saw this basic process as the force driving a grand cycle of regeneration in which the sea and land continually changed places—continents eroding into oceans to form great piles of sediment that eventually melted at the base and rose anew.

Hutton's ability to imagine an endless cycle of erosion and deposition that led to the formation of fresh rocks kicked open the door for serious consideration of the immensity of geologic time. He wasn't arguing that the world was older than imagined; he flat-out argued that Earth was ancient beyond imagination. Who could know how many times rocks had been recycled? There was no way to know how many cycles of erosion and uplift the world had seen when each cycle destroyed evidence of prior ones. He must have shocked an audience that believed Werner's ideas about rocks precipitating out of the ocean on a not quite six-thousand-year-old world. His extreme views even startled those willing to consider more expansive views of geologic time, like Buffon's. The skeptical reception of those present spurred him to seek out more evidence to bolster his arguments.

When his lecture was finally published three years later, in 1788, it garnered dismissive reviews that mischaracterized his unknowably old Earth as a rewarmed version of Aristotle's eternal world without beginning or end. Particularly controversial was the contention that the world evolved in a cyclical fashion. This was totally at odds with the Mosaic account in Genesis of the Creation and the Flood. Everyone knew that things hadn't happened over and over again. The idea that new land was pushed up from beneath the sea by the force of Earth's internal heat placed Hutton squarely at odds with both Werner's Neptunist disciples, who believed in an aqueous origin for rocks, and the traditional Christian conception of a recently created, decaying world.

A simple test of Hutton's idea lay in determining whether granite veins formed along with or were younger than the rocks they were found in. If precipitated together from an ancient sea, rocks and the veins they harbored should be the same age. If Hutton was right—that molten rock rose up from deep below the seafloor—then the veins should cut across the sedimentary layers.

Scouring the highlands on field excursions, Hutton sought out layered rock cut by veins of granite. He found what he was looking for in the boulders and exposed bedrock riverbed of idyllic Glen Tilt, a valley west of Aberdeen. There veins of red granite clearly passed through bed after bed of black sedimentary rock. The granite had intruded the sedimentary rock after it was formed. The thin stripes of granite were indeed younger than the rocks in which they were found.

The following summer, Hutton found more granite veins injected into sedimentary rocks in Galloway, in southwest Scotland. Even better than at Glen Tilt, these veins terminated within the exposed strata, only penetrating partway up into the stacked sediments. Not only was the granite younger, it came from below. Here was more evidence that granite did not precipitate out of an ancient sea. Hutton felt increasingly confident that what he was seeing revealed that Earth was far older than anyone believed.

This wasn't enough to prove Hutton's grand cycle. It only validated his mechanism for uplifting rocks through heat from below. Confident he was right about the larger story, he kept looking. Three years after he boldly announced that the world was immeasurably old, he sailed south from Edinburgh, searching the North Sea coastline for outcrops that would support his ideas. Two colleagues joined him: John Playfair, a professor of mathematics at the University of Edinburgh, and twenty-seven-year-old Sir James Hall, grandnephew of the influential president of the Royal Society. Playfair was a former Presbyterian minister steeped in the traditional views of the Scottish church. Hall, a wealthy young man, supplied a boat and crew for the day, allowing their party to cover far more ground that they could on foot. Both Playfair and Hall had initially rejected Hutton's idea of an ancient Earth. Now, after years of discussions, Hutton had begun to convince them that he might be on to something.

Hutton picked this stretch of coast to explore because he knew the area was composed of two types of rock—fine-grained gray sandstone and coarser red sandstone. He was convinced that these strikingly different rocks represented two distinct cycles of uplift and sedimentation. Somewhere along the coast the two formations would meet, and the eroding sea cliffs could expose their contact.

They sailed south from Hall's estate along the rocky coast, where high cliffs provided excellent exposures of the older gray rock. Several headlands down, they passed a sandy beach where the beds in a red sandstone cliff lay pitched at a twenty-degree angle. But where did the red rock meet gray rock? Around the next headland they struck gold. At the base of the cliff, vertical layers of gray rock jutted upward only to encounter the overlying red sandstone. In between the two rock formations lay gray rubble that looked like the modern beach deposits exposed along the shoreline.

Hutton was ecstatic. The contact between the gray and red rock lay exposed in striking clarity, and the story it told demolished conventional views. Here was proof of several rounds of Hutton's grand cycles. Playfair later described the moment in dizzying terms that evoke a religious epiphany.

> *The mind seemed to grow giddy by looking so far into the abyss of time; and while we listened with earnestness and admiration to the philosopher who was now unfolding these wonderful events, we became sensible how much farther reason may sometimes go than imagination can venture to follow.*[2]

Hutton had converted his field companions, but who else would believe that the world was unbelievably old? Who else would dare to imagine that cycles operating at a planetary scale could explain the origin of rocks and ultimately the world we see around us?

When Hutton published his *Theory of the Earth* in 1795, Werner's Neptunism dominated geological opinion. Hutton's near-vertical layers of once-horizontal secondary rock, by then widely acknowledged to predate the Flood, told of far more than Burnet's collapsing crust or Werner's gradually drying oceans. Hutton argued that mountains and oceans traded places over and over again in a global dance of erosion and sedimentation that demonstrated a divine design. He rejected the role of catastrophes like the biblical flood not only because they ran counter to his own observations but because periodic destruction of the world ran counter to his view of a divine design to everything on Earth. Faith in the perfection of God's principles favored slow geological change—uniformity of action rather than violent catastrophes. Hutton saw Earth as a grand machine set in motion by natural laws that ran a perpetually self-renewing system he famously characterized as having "no vestige of a beginning,—no prospect of an end."[3]

His peers thought him crazy.

To some degree, the cool reception of Hutton's ideas reflected the politics of his time. Upper-class British intellectuals shocked by the excesses of revolutionary France saw a rising tide of atheism as fueling the horrors of the guillotine. Hutton's rejection of both conventional biblical chronology and Noah's Flood as the driving force of geologic history placed him in league with radicals set on overthrowing civilization. It hardly mattered that Hutton himself was deeply conservative. His ideas about an ancient Earth challenged tradition and authority.

Still, the rocks at Siccar Point simply did not fit into the model of a global flood as the singular event in earth history. The rocks were evidence of two geological eras separated by an abyss of time. Any way one looked at it, the eons necessary to explain the cycle of worlds apparent in Hutton's two rounds of uplift and erosion did not fit with a literal reading of Genesis.

Hutton's critics were not easily deterred. In 1793 one of Werner's students, Richard Kirwan, savaged Hutton's theory in the *Transactions of the Royal Irish Academy*, essentially accusing him of being an atheist. Hutton immediately began working on a greatly expanded version of his theory that would show how God established the world's geological order at an unknowable date in the distant past and would terminate it at some unknowable date in the future. Just when the world began and when it would end were metaphysical issues beyond the reach of rational inquiry.

While frantically working to reframe and support his case, Hutton contracted a debilitating illness from which he never recovered. He completed two of three planned volumes of his *Theory of the Earth* despite great pain, which goes a long way toward explaining why the book is famously unreadable. Hutton died in March 1797, shortly after scathing reviews once again dismissed his ancient planet theory as a warmed-over version of Aristotle's pagan eternal world.

Hutton's Irish nemesis kept at it, marshaling geological evidence to defend Werner's Neptunism against Hutton's heat-driven theory and its heretically ceaseless cycles of uplift and erosion. Published in 1799, Kirwan's *Geological Essays* attacked Hutton's theory on moral and religious grounds. Kirwan thought the idea of an ancient Earth undermined society's foundation: "how fatal the suspicion of the high antiquity of the globe has been to the credit of Mosaic history, and consequently to religion and morality."[4] Kirwan found Hutton's arguments so absurd that in preparing his rebuttal he reportedly didn't even bother to read the Scotsman's book.

Instead, like others before him, he came up with another novel theory to explain Noah's Flood. As Kirwan's primitive Earth precipitated from primordial fluid, the water level gradually sank to that of the present oceans, leaving the continents high and dry. Misinterpreting frozen mammoth remains as drowned African elephants, Kirwan proposed a new idea to explain how their bones got to northern Europe and Siberia. In the beginning, long before the Flood, a globe-covering sea gradually retreated down into great rifts in Earth's crust. Much later, all that water was released suddenly, triggering Noah's Flood somewhere between India and the South Pole and transporting the remains of tropical animals to Siberia. No Northern Hemisphere creatures were found in the Southern Hemisphere, but elephants (mammoths) kept turning up in gravel deposits at high latitudes. Unaware that these enormous carcasses were almost always solitary (and quite hairy), Kirwan imagined that great piles of elephant bones showed how the beasts huddled together to face the oncoming flood before they were swept off to Siberia. He ignored the puzzling lack of lion, zebra, giraffe, and other bones of African animals in the Siberian deposits.

In Kirwan's mind, floodwaters racing north reshaped continents, destroying an ancient landmass between Asia and North America and leaving Mongolia's Gobi Desert a vast barren flat. He didn't stop there, explaining how the Flood turned the Arabian Peninsula and North Africa into wasteland, and carved out the Bay of Bengal and the Red and Caspian seas. The planet's shattered crust kept settling and producing earthquakes until around 2000 BC, creating Gibraltar, the Dardanelles, and the Straits of Dover. Putrefying remains of plants and animals sucked enough oxygen out of the atmosphere to reduce humanity to its present enfeebled state. And because carnivores would have been hard to manage on the ark, Kirwan proposed that God recreated them all after the Flood, along with the entire American fauna. He liked this idea so much he didn't mind that the Bible neglected to mention this second round of creation.

Although Kirwan was fervent in his desire to defend the traditional literal interpretation of Noah's Flood, he abandoned biblical literalism to bring in additional details and events not described in the Bible. He made up a geological story to preserve his preferred reading of the biblical story. Others, however, began to accept an older Earth in attempts to harmonize the biblical flood with the story told by the rocks.

Hutton's influence would withstand the test of time, but he never had a chance to respond to Kirwan. Hutton's impenetrable book was not compelling enough to convert skeptics to his side. Lacking the planned third volume that was to have related his discoveries at Glen Tilt and Siccar Point, his *Theory of the Earth* nearly died with him. Playfair and Hall, the primary witnesses to his field excursions, would not let Hutton's work languish. They began spreading Hutton's gospel of deep time. When sorting through his colleague's papers while writing a memorial, Playfair realized just how much persuasive material Hutton had been working up for his unpublished third volume.

Determined to promote his friend's ideas, Playfair did what Hutton had longed and meant to do and completed a compelling treatise about the antiquity of the Earth. He graciously credited his late friend by titling his work, published in 1802, *Illustrations of the Huttonian Theory of the Earth*. Here was an impressive, engaging work that included a distillation of Hutton's theory followed by elaborations, examples, and responses to criticisms, much as Hutton himself had dreamed of presenting it.

In bringing Hutton's ideas to the attention of mainstream scholars, Playfair also explained how rivers could carve topography—given enough time. He argued that "rivers have, in general, hollowed out their valleys" because "the general structure of valleys among mountains is highly unfavorable to the notion that they were produced by any single great torrent, which swept over the surface of the earth."[5] Playfair saw that valleys diverged in all directions from the center of mountain ranges, so a single current sweeping across the terrain could not have carved them all. Neither could a single current have carved valleys running at right angles to each other or perpendicular to the overall trend of the drainage from a mountain range. He went on to describe the division of landscapes into integrated networks of little valleys connected to larger valleys, each seamlessly connecting with the next at a common elevation regardless of size. Such a landscape could only be the signature of running water slowly eroding Earth's surface. Ahead of his time by decades, Playfair made a compelling case that Noah's Flood did not shape the world's topography.

Playfair also addressed Siberian mammoths. He noted that their bones were always found in soil or alluvial deposits and never in the solid rock below. Writing in the style of his time, he rambled on a bit before pointedly demolishing Kirwan's conception of the Flood.

> *If we consider attentively the facts that respect the Siberian fossil bones, there will appear insurmountable objections to every theory that supposes them to be exotic, and to have been brought into their*

> *present situation from a distant country. . . . Shall we ascribe it to some immense torrent, which, sweeping across the desarts of Tartary, and the mountains of Altai, transported the productions of India to the plains of Siberia, and interred in the mud of the Lena animals that had fed on the banks of the Barampooter or the Ganges? Were all other objections of so extraordinary a supposition removed, the preservation of the hide and muscles of a dead animal, and the adhesion of the parts, while it was dragged for 2000 miles over some of the highest and most rugged mountains in the world, is too absurd to be for a moment admitted.*[6]

Playfair further noted how their carcasses would surely have rotted if these great beasts had died in a tropical climate. Whatever they were, mammoths were not relics of the Flood.

By the close of the eighteenth century, theologians had begun to recognize the lack of a unified explanation among natural philosophers for Noah's Flood and the age of the world. The wide range of conflicting theories and interpretations fostered suspicions that perhaps it was the Bible that was being misinterpreted. The floodgates of heaven and the fountains of the deep had been interpreted to refer to comets, a great vapor canopy, water from alpine caves, and a vast subterranean sea—just about everywhere one might imagine finding enough water to drown the world. Theologians started to question whether scripture was meant to be a source of scientific information as well as a book of personal and moral redemption. Even conservative Christians began to question whether Noah's Flood was all there was to earth history.

It is impossible to stand at Siccar Point and reasonably see how to fit what you can read in the rocks into just 6,000 years of time. When Roman ruins still stand after 2,000 years, how could raising and eroding off two mountain ranges happen in just twice as long before that? The virtually unimaginable amount of time required to form the two unconformable sandstones exposed along the Scottish coast offers a humbling glimpse of the infinite.

Hutton's recognition of the concept we now call deep time laid the foundation for a new geological time scale. It was a turning point in our story and a huge development for the field of geology. Reinterpreting the days of the week of Creation as geological ages allowed earth history to accommodate vast expanses of geologic time. After all, who knew how long one of God's working days lasted? Perhaps the rock record paralleled Genesis—if interpreted as consisting of six ages rather than six days. Maybe Moses only wrote about the part of the Flood that Noah witnessed. Although biblical interpretations were being reconsidered, there was still general faith that the rocks filled in the real story.

Then, as now, conventional wisdom guided interpretation of discoveries to the extent it could. Scientific revolutions happen when conventional views can no longer bend under the weight of new findings. Natural philosophers were still looking to prove Noah's Flood because they viewed the world through the filter of religion, not because they feared theological condemnation. Despite the evidence Hutton and company marshaled to frame the geological story, natural philosophers were reluctant to abandon the biblical story. Only later did science start to modify and seriously undermine faith in biblical truth. Even so, it had become clear there was more to earth history and fossils than simple deposition of sedimentary rocks from a single flood over the span of a single year.

Soon geologists would unearth compelling evidence for multiple catastrophes, each of which ended a distinct period of earth history. As nagging questions and alternative ideas began to reshape how Christians interpreted the story of Noah's Flood, natural philosophers shifted gears in looking for geological evidence of it. The search for Noah's Flood moved from rocks into the overlying deposits of unconsolidated sediments that lay scattered across Earth's surface.

7

Catastrophic Revelations

BEFORE THE EARLY NINETEENTH CENTURY, natural philosophers paid little attention to deposits of loose gravel, sand, and boulders lying above solid rock. But northern Europe's geological blanket of unconsolidated material became far more interesting once it was thought that the part of earth history that overlapped with human history was preserved in surficial sediments rather than in the solid rock below. It helped that geology arose as a science in countries that had been glaciated, where a regional cover of glacial deposits—gravel, sand, boulders, and mud—resembled what you might expect a big flood to leave behind. These surface deposits and topography, the form of the land itself, became the link between the modern world people knew and the former worlds preserved in the rocky depths of geological time.

I came to appreciate the potential for catastrophic rearrangement of surficial deposits in the Philippines. At the time, I was doing fieldwork in the Pasig-Potrero River, where one of my graduate students was studying changes after the catastrophic 1991 eruption blew the top off Mount Pinatubo and buried the surrounding countryside under hot pumice and ash. The whole landscape around the volcano changed, as river valleys filled in with sediment only to have great canyons cut back down hundreds of feet into the loose debris in just a couple of years. We saw the Passig-Potrero River as an ideal place to study how rivers behaved when supplied with as much sediment as they could carry.

On a beautiful tropical morning, we started out from Delta 5, an abandoned military checkpoint perched on a rock outcrop sticking up from the riverbank. We headed upstream, leaving the coastal plains to enter the volcanic upland. Walking up the riverbed, we surveyed it in three-hundred-foot sections. One person would stay behind, sighting through a tripod-mounted level, as another took our stadia rod—a giant collapsible ruler—out to the end of a long tape measure. Using the level to read off the elevation every few feet as we moved the stadia rod along the tape, we measured the elevation of the riverbed. Repeating the survey over a number of years gave us a record of how the river ate down into the volcanic debris as lahars—volcanic mudflows—surged downstream to bury villages and towns beneath a blanket of sediment.

Just before lunch we noticed that an ominous black cloud had settled in over the volcano several miles upstream. The river started rising as we kept surveying our way up through a tight canyon. When the flow got deep enough to start moving the riverbed, grapefruit-sized rocks rolled into our shins and we decided to break for lunch on a sand terrace five or six feet above the water level. About halfway through lunch we noticed the water rising even faster. As the river started lapping up onto our lunch-stop terrace, we retreated to the foot of the canyon walls and watched six-foot-high waves cascade down the river we had walked up all morning.

Alarmed, we climbed up through narrow side channels that had cut down through the volcanic debris—the only other way out of the

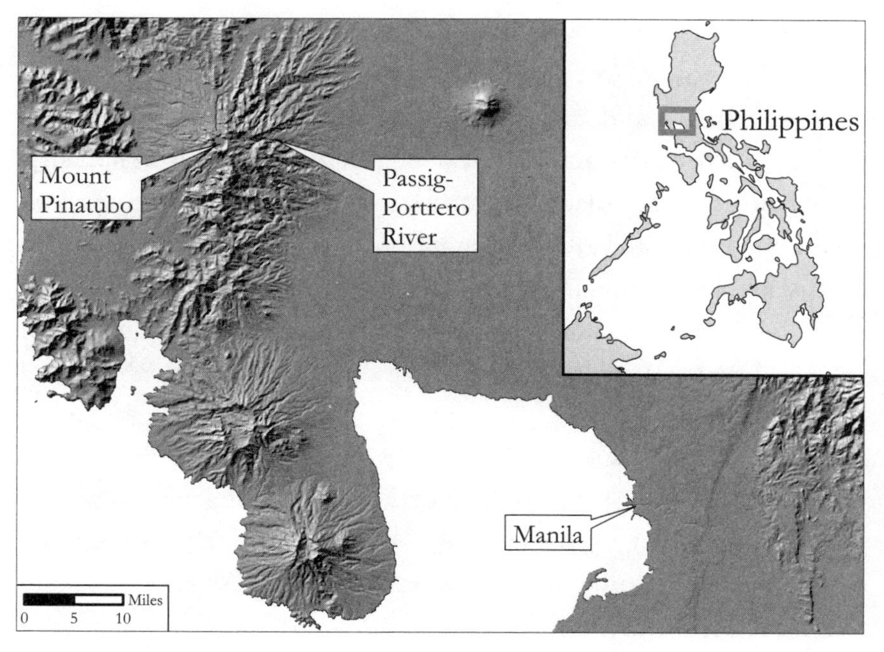

Map of the Philippines showing location of Mount Pinatubo and the Pasig-Potrero River draining off the volcano's eastern flank.

canyon. By the time we reached the top of the side canyon we could see our lunch spot, several hundred feet below, submerged beneath a roaring torrent of bouncing boulders. We perched for the afternoon, trapped on the ridgetop but enthralled by walls of water crashing down the canyon. Here in front of us was a graphic illustration of what had drawn me to geology in the first place: Earth's appearance of being stable—of being solid as a rock—only held some of the time.

In the early nineteenth century, the conventional view was that grand catastrophes reshaped landscapes in a geologic jiffy. The idea that the slow pace of everyday change could shape and reshape the world was considered delusional fantasy. By the end of the century, geologists believed that everyday erosion was how the world worked, and grand catastrophes had become geologically taboo.

Scientific curiosity and religious conviction were not alone in pushing efforts to better understand regional geology. Just as demand for iron and coal drove advances in mining and mineralogy, construction of railroads and canals created a need to understand regional geology. As necessity and practical interest grew, schools in industrializing areas began to appoint professors of geology. Studying rocks could be more than just an inspired hobby for those with the time, means, and inclination to seek insight into nature's inner workings. It could be a livelihood. As geologists began to work out the details of local and regional geology, they reassessed the role of Noah's Flood in earth history.

In 1815, surveyor and canal builder William Smith worked out the structure of England's layered rocks in compiling what is widely credited as the first regional geologic map. He carefully documented a consistent, well-ordered succession of rock types across England that was far too systematic to have formed during the chaos of a globe-wrecking deluge. Smith also showed that different layers of rock consistently held different fossils. Based on observations collected over years of field excursions, Smith's carefully compiled map allowed him to accurately predict the type of rock and the fossils in it virtually anywhere in England. His obsession with perfecting his map bankrupted both himself and the idea that a single catastrophic flood deposited layered rocks. After he published his map, geologists no longer looked for Noah's Flood in the rocks. Instead they looked for signs of a great flood in topography and surficial deposits.

Across the English Channel, Smith's contemporary Georges Cuvier, the vertebrate paleontologist who had dismissed Scheuchzer's flood victim and concluded that mammoths were extinct, was busy mapping the rocks in the countryside around Paris. He found a sequence of distinctively terrestrial rocks containing fossil quadrupeds that alternated with layers full of fossil seashells. He knew that a single flood could not produce a thick sequence of interlayered terrestrial and marine rocks. Clearly, the sea inundated the land not just once but time and time again. Further fieldwork in the Paris basin unearthed evidence for alternating periods of fresh and salt-water inundation that Cuvier interpreted as evidence for at least half a dozen great floods, each of which ended a geological era. Instead of Hutton's grand engine of slow change, Cuvier's 1813 *Essay on the Theory of the Earth* concluded that each catastrophe recorded another transition in a long series of geological eras. Ever since, these two views of geologic change—slow and steady versus catastrophic—have framed competing theories for how the world is shaped.

The idea that a catastrophic biblical flood could have remodeled the European landscape was vividly reinforced in 1818, when the Getroz glacier dammed the river Dranse in Switzerland's Val de Bagnes. Advancing like the glacier that dammed the Tsangpo in Tibet, the ice blocked the river and a lake holding eight hundred million cubic feet of water formed above the frozen impoundment. When a tunnel was cut through it to draw down the lake, the ice and debris dam failed, sending a wall of debris-charged water surging down the valley at more than thirty feet a second. The flood swept away landmarks as sand and mud filled the local church to the pulpit. Huge boulders lay strewn around the fresh deposits. As residents dug out from the mess, they discovered trees and houses swept away in the torrent. The event impressed natural philosophers with how catastrophes could blanket large areas under sediment. Here, perhaps, was an analog for the geological signature of really big floods. The deposit left by this modern catastrophe looked a lot like the blanket of sand, gravel, and mud that covered much of northern Europe.

Again, Cuvier led the way in elaborating the power and dynamism of geological processes in his 1825 *Discourse on the Revolutions of the Globe*. He made the case that distinctive animals lived during different epochs of earth history and described how abrupt discontinuities between geological formations with different fossil assemblages testified to periodic catastrophes having remodeled the world. In his view, the most recent catastrophe was a sudden flood that separated the relatively short history of humanity from the depths of geologic time. Cuvier's contention that one could not explain the geologic record solely by means of existing causes—that the processes that shaped Earth's surface were different in the past—became known as catastrophism, and stood in direct contrast to Hutton's articulation of how things happened gradually through many small changes, a view that became known as uniformitarianism.

Cuvier's idea of periodic cataclysms seemed to address otherwise perplexing observations. His compelling evidence for the repeated destruction of former worlds inspired geologically literate clergy to reinterpret Genesis. As early as 1816 the Stackhouse Bible cautioned readers, "Moses records the history of the earth only in its present state. . . . There is nothing in the sacred writing forbidding us to suppose that [fossils] are the ruins of a former earth."[1] Fossils now belonged to numerous ancient catastrophes. Geological evidence was starting to shape biblical interpretation.

A prominent Protestant, Cuvier did little to counter the impression that the most recent of his long series of grand catastrophes was the biblical flood. He asserted it could not have been all that ancient: "If there is any circumstance thoroughly established in geology, it is that the crust of our globe has been subjected to a great and sudden revolution, the epoch of which cannot be dated much farther back than five or six thousand years ago."[2] He thought that a small number of people and animals survived the most recent cataclysm, about the time conventionally ascribed to Noah's Flood.

Those seeking geological support for the biblical flood now looked to the sediments on top of the rocks, assuming Noah's Flood was a more recent catastrophe than the geological revolutions recorded in hard rock. The most influential nineteenth-century diluvialist was William Buckland, a minister in the Church of England and

Oxford's first professor of geology. He passionately defended the traditional view of Noah's Flood but acknowledged that the six days of Creation could not be taken literally. The son of a clergyman, Buckland knew that geology would instantly become a respectable science if he could show that it validated the Genesis flood.

A man of his times, Buckland straddled both worlds—those of the church and field geology. He wanted to forge links between human history as recorded in classical texts and biblical stories and earth history as revealed by geology. Like many of his contemporaries, he believed that Moses disregarded most of earth history because it did not concern mankind.

Confident of the reality of Noah's Flood, Buckland saw its signature in the sculpting of topography and the geologically recent deposition of the blanket of gravel covering much of Britain. He saw geological evidence as supporting the universality of the Deluge. What else could explain the giant out-of-place boulders in northern Europe from Norway to the Alps? Made of rock with no local source, boulders the size of barns had obviously been transported from distant sources. A really big flood seemed like the only reasonable way to explain how to move huge rocks. Lacking reasonable alternatives, Buckland and his contemporaries attributed the deposition of the gravel blanket and transport of enormous boulders to great waves during the biblical flood.

In his 1819 inaugural address at Oxford, Buckland equated Cuvier's most recent catastrophic inundation with Noah's Flood.

> *The grand fact of an universal deluge at no very remote period is proved on grounds so decisive and incontrovertible, that, had we never heard of such an event from Scripture, or any other, authority, Geology of itself must have called in the assistance of some such catastrophe, to explain the phenomena of diluvian action which are universally presented to us, and which are unintelligible without*

recourse to a deluge exerting its ravages at a period not more ancient than that announced in the Book of Genesis.[3]

Although the remains of modern species buried in the surficial gravels pointed to a recent calamity, Buckland did not believe that Noah's Flood formed fossil-bearing rocks. To find evidence of the Flood you had to look in the overlying unconsolidated sediments and at the lay of the land, the form of topography.

In Buckland's opinion, Europe's surficial gravel was too extensive to have been laid down by rivers. He thought the Flood simultaneously deposited it and carved the modern landscape from older rocks. Buckland coined the term diluvium to describe the surficial deposits that mantled much of northern Europe and to distinguish them from alluvium, the sand and gravel laid down by modern rivers. He remained disturbed, however, that no human fossils had been found in diluvium. Where were the bones of those the Flood was sent to destroy?

Despite this troubling detail, Buckland stressed that geological facts were broadly consistent with the biblical account because Noah's Flood ushered in only the most recent of a long succession of worlds. Buckland's lecture, published as *Vindiciae Geologicae; or, the Connexion of Geology with Religion Explained*, argued that geological facts "are consistent with the accounts of the creation and deluge recorded in the mosaic writings. . . . The evidences afforded by Geological phenomena may enable us to lay more securely the very foundations of Natural Theology."[4]

The "Natural Theology" to which Buckland referred followed William Paley's popular and influential 1802 book of the same name. Paley argued that scientific revelations contradicting biblical interpretations provided natural guidance for better interpreting scripture because the Bible and the book of nature shared the same author. In the opening decades of the nineteenth century, even Pope Pius VII endorsed viewing the six days of Creation as of indeterminate length rather than as a literal week of twenty-four-hour days. A little more than a decade after publication of Paley's popular book, in 1813, English geologist Robert Bakewell sought to reconcile the geological and biblical chronologies in his *Introduction to Geology*, the first geological textbook published in English, arguing that the Mosaic chronology began when the world became fit for human habitation.

Others argued that a long time passed between the initial Creation in the first verse of Genesis and the formless Earth of the second verse. Perhaps the time between when God created the world long ago and when he remodeled it for human use wasn't recorded in the Bible, leaving an indeterminate gap between the first two verses of Genesis. The gap theory, as this idea became known, provided an alternative to the day-age theory that each day of creation lasted far longer than twenty-four hours.

Two centuries ago, Christian scholars adapted how they read the Bible to account for geological revelations. And why not? The history of the world that geologists had found in the rocks followed the order of events described in Genesis—an initial period of time without life, followed by the introduction of plants and animals, and eventually people. If the days of Creation referred not to a single week of breakneck change but to a long series of geological ages, the problem that more than six days was needed to account for prehistory became an interpretive detail that did not imperil scriptural authority. Nowhere, Buckland asserted, did Genesis contradict the idea that the modern world was built upon the ruins of prehuman worlds. With one foot in the newborn profession of geology and the other in Anglican orthodoxy, Buckland was a man of deep conviction and few doubts.

Most geologists love the field aspect of our work, and Buckland appears no different. He went on field excursions across Britain and Europe, accompanying natural philosophers he visited and in the company of those visiting him. He traced the occurrence of durably hard yet smoothly rounded quartzite pebbles in surficial gravels from Oxford north to Warwickshire. There, he found these distinctive pebbles eroding from outcrops of conglomerate, rock formed when gravel and sand were buried deep enough to turn back into solid rock. This unusual formation was known as pudding stone due to the resemblance of the gravel set in a sand matrix to plums in a Christmas pudding. Through his geological sleuthing, Buckland reasoned that the quartzite pebbles had to have been rounded before being incorporated into the conglomerate. He thought that a great flood then ripped the distinctive pebbles back out of the rock, strewing them down the Thames all the way to London.

Buckland claimed that a great flood provided a better explanation for the distribution of the diluvial gravels than did other ideas—modern rivers were too small to account for regionally extensive gravel sheets or to move the largest boulders found in the deposits. And what at the time seemed like an apparently global distribution of similar deposits was thought to demonstrate that a geologically recent flood had affected the surface of the entire world. Again, Buckland was confident that a great flood provided the best explanation for his geological observations.

It should come as no surprise, then, that he marveled over what he considered proof of Noah's Flood when workmen in 1821 discovered a bone-filled cave near Kirkdale in Yorkshire. One of the first to explore the cavern, Buckland stumbled upon a bewildering variety of bones, including those of hyenas, tigers, elephants, rhinoceroses, and hippopotamuses. All these bones were embedded beneath stalactites in the red mud of the cave floor. It was a spectacular discovery indeed.

How did the bones of so many African species get mixed up together in a British cave? Seeing how some of the bones were gnawed, Buckland concluded hyenas had dragged them into their den long before the Flood, which he thought washed in the cave's uppermost layer of red mud and more bones. The thin stalactites capping the mud confirmed a recent origin, consistent with Cuvier's most recent geological catastrophe of five or six thousand years ago.

Inspired, Buckland gathered geological facts thought to demonstrate the reality of Noah's Flood into his 1823 *Relics of the Flood.* In it he described great accumulations of bones in "superficial and almost universal deposits of loam and gravel, which seems impossible to account for unless we ascribe them to a transient deluge, affecting universally, simultaneously, and at no very distant period, the entire surface of our planet."[5] The case for Noah's Flood appeared to build once again, this time in the interpretation of surficial sediments.

Buckland combined his description of Kirkdale Cave with a synopsis of similar evidence for a recent deluge from other European caves. The continent's surficial gravel contained exotic fossils like those from Kirkdale Cave and unlike modern species. Other evidence included giant blocks of granite from Mont Blanc scattered well beyond the Alps. Rejecting a southern origin for the Flood, he argued that Europe's surficial gravel and stray boulders came from identifiable northerly sources. He also maintained that the violent floodwaters carved valleys far too deep and wide to have been cut by the piddling rivers that flowed through them today.

In coming to these conclusions, Buckland relied on what he saw with his own eyes. Nowhere did he invoke scriptural authority, even if it framed his view. His reasoning was compelling enough that others hailed his explanation as vindication for the reality of Noah's Flood. Like Cuvier, he did nothing to discourage the idea. After all, his defense of a global flood had its rewards. Even before his work on Kirkdale Cave, Buckland received the Royal Society's prestigious Copley Medal. Appointed Canon of Oxford's Christchurch Cathedral three years later, he eventually became Dean of Westminster, one of the most prestigious positions in the Anglican Church.

Buckland was hardly alone in thinking he had found evidence of Noah's Flood. Adam Sedgwick, who held Woodward's old chair as professor of geology at Cambridge and taught Darwin his geology, summarized conventional thinking in 1825.

> *The sacred records tell us—that a few thousand years ago 'the foundations of the great deep' were broken up—and that the earth's surface was submerged by the water of a general deluge . . . [which] has left traces of its operation in the diluvial detritus which is spread out over all the strata of the world.*[6]

Not long afterward, cracks began developing in Buckland's geological case for a global flood.

The end began when flood skeptics like John Fleming, an evangelical pastor in the Church of Scotland and professor of natural philosophy at Aberdeen, questioned the arguments and conclusions of flood champions like Cuvier and Buckland on theological as well as geological grounds. Fleming's 1826 article in the *Edinburgh Philosophical Journal* used logic and literal interpretations of scripture to challenge Buckland's version of the Flood.

Fleming opened with the problem of how Buckland could attribute extinctions to the Flood when the Bible said that two of every creature boarded the ark. If Noah saved a pair of all the world's animals, then geologists could not blame extinctions on the Flood. And the biblical flood sounded like a relatively tranquil affair, leaving submerged olive trees intact after taking forty days and nights for the waters to rise. To Fleming, a literal interpretation of the biblical story was inconsistent with Buckland's view of violent currents capable of carving deep valleys into hard rock and transporting huge boulders and carcasses halfway around the world. Fleming granted that a great flood could have swept away loose soil but doubted that so brief an event could have gouged out deep valleys. To the contrary, a literal reading of Genesis implied that the ark grounded out close to where Noah and his crew first embarked. Surely a flood powerful enough to reshape the world would strand Noah somewhere far from where he started.

Although Fleming made it clear that he did not question the occurrence of the biblical flood, he viewed the affair as tranquil enough to leave no geological signature. He considered it futile to look for physical evidence of the Flood.

Fleming also questioned Buckland's geological interpretations. A global flood would leave the same kind of mud in caves all across Europe. Yet the mud one found varied with the local geology. And if the mud wasn't washed in from afar, how could the fossils entombed in it have been?

Fleming's critique continued with summarily dismissing the theory that the elephantlike bones and carcasses found in Siberia and North America came from tropical regions. The intact skeletons ruled out long-distance transport by a violent deluge. Pointing to Cuvier's anatomical studies, Fleming argued that the thick hair covering mammoth carcasses showed they were native to cold regions. These behemoths were well suited to living where their bodies were found. Mammoths did not confirm the transporting power of the Flood.

Fleming even questioned Buckland's interpretation of Kirkdale Cave. While he agreed that the cave was an ancient hyena den, he thought that Buckland jumped to conclusions in attributing to a single flood the mud in which the bones were found. A succession of small floods could have deposited the mud.

Reverend Fleming chided geologists for rushing to find evidence of the biblical flood. In his view, misguided efforts to use geology to vindicate biblical interpretations would harm both science and Christianity.

More than Fleming's scathing critique, new geological discoveries eroded Buckland's faith in a universal deluge. Most problematic for a global flood was that explorers could find no diluvium in the tropics. Closer to home, it proved impossible to explain the complex stratigraphy of European diluvium through a single event, no matter how catastrophic. Buckland began to reconsider whether his imagination had run wild in his zeal to defend the biblical flood. A decade after Fleming first challenged him, Buckland capitulated when he was asked to prepare a volume commissioned by the estate of the Earl of Bridgewater to illustrate how geology revealed the wonder and wisdom of Creation.

In 1836, Buckland did something few others before him had done in attempts to reconcile geology and the Bible. He pulled a complete about-face when his Bridgewater volume *Geology and Mineralogy* repudiated his earlier view of diluvium. Instead, he endorsed the position that a tranquil Flood did little to Earth's surface, long after earlier catastrophes laid down fossil-bearing rocks and surficial deposits. Citing recent discoveries, Buckland advocated caution in trying to use the geological record to support literal interpretations of Genesis.

> *The disappointment of those who look for a detailed account of geological phenomena in the Bible, rests on a gratuitous expectation of finding therein historical information, respecting all the operations of the Creator in times and places with which the human race has no concern; . . . the history of geological phenomena . . . may be fit matter for an encyclopedia of science, but are foreign to . . . a volume intended only to be a guide of religious belief and moral conduct.*[7]

Although Buckland still maintained that a geologically recent inundation overwhelmed the northern hemisphere, his earlier confidence that it was the biblical flood lay shattered. He could no longer attribute fossils to Noah's Flood. Fossils were found in strata that accumulated slowly over long periods of time. Even the surficial deposits recorded more than one event. Buckland had abandoned Noah's Flood.

Despite his change of mind, Buckland had no concern that geology and revelation would prove inconsistent.

> *Geology has shared the fate of other infant sciences in being for a while considered hostile to revealed religion; but, when fully understood, it will be found a potent and consistent auxiliary to it, exalting our conviction of the Power, and Wisdom, and Goodness of the Creator.*[8]

Secure as ever in his faith in both nature and the Bible, Buckland maintained that the question is not "the correctness of the Mosaic narrative, but of our interpretation of it."[9] In a philosophical turnabout, Buckland shifted from using geology to shore up a literal interpretation of the Bible to arguing that biblical interpretations could be tested through consistency with geological observations.

Coming from a conservative man of the cloth, Buckland's Bridgewater treatise drew immediate attacks from fellow clergy appalled by his recantation of geological support for the biblical flood. Outraged traditionalists who insisted on interpreting the Bible literally railed against this compelling dismissal of scriptural geology by a ranking clergyman steeped in Anglican orthodoxy.

What led to Buckland's stunning reversal? To a great degree it was his former spellbound student, Charles Lyell.

Born into a life of privilege the year James Hutton died, Lyell grew up exploring the New Forest in Hampshire, where his father pursued botanical studies and encouraged his son's interest in the family's extensive natural history library. Raised an Anglican, Lyell read Bakewell's just-published geology textbook in 1816, the year he enrolled at Oxford to study classical literature and law. Lyell was particularly struck by Bakewell's concept of a world much older than generally supposed based on a literal reading of Genesis. Equally intriguing to him, this unconventional idea came from the pen of someone who believed geology revealed the Creator's grand design.

Lyell attended Buckland's Oxford lectures each spring from 1817 to 1819. He came to accept that the biblical chronology referred to the time since the creation of people. Who could know how much time had passed before then?

Buckland's enthusiastic endorsement ensured Lyell membership in the Geological Society of London once he graduated. Society members overwhelmingly rejected Hutton's view of great cycles of gradual change driven by processes like those operating at present. Most advocated Cuvier's view of earth history as a series of violent catastrophes. On a visit to Paris the previous year, Lyell examined Cuvier's collection of fossils, describing them as "glorious relics of a former world."[10]

After graduation, Lyell divided his time between reading for the bar and traveling through Europe. Visiting Paris again in 1823 as a representative of the Geological Society, he met Constant Prévost, a colleague of Cuvier, who believed that the alternating freshwater and marine strata of the Paris basin were deposited gradually in a coastal inlet that periodically turned into a freshwater lake. Perhaps geological change could occur through observable causes, if given enough time.

The following year, in the fall of 1824, Lyell visited sixty-three-year-old James Hall at his estate on the Scottish coast. Now about the age Hutton was when they first sailed to Siccar Point, Hall took Lyell there to absorb Hutton's insight through his own eyes. Seeing firsthand how earth history involved a lot more time than conventionally thought, Lyell began to believe that gradual changes could shape the land.

That same year, Lyell joined Buckland for a geological excursion through Scotland. Although it may have been clear to both that their views had started diverging, neither could have known that within a decade the apprentice would dethrone the master.

Lyell was not particularly interested in questioning religious views. Like many of his peers, however, he was deeply concerned about the effect that ignoring geological evidence could have on both science and religion. In 1827, he concluded a review of George Poulett Scrope's *Memoir on the Geology of Central France* with an appeal for interpreting Genesis broadly and letting the rocks speak for themselves:

> *We must recollect that the Mosaic narration is elliptical in the extreme, and that it makes no pretensions whatever to supply those minute scientific details which some would endeavour to extort from it.*[11]

Lyell was echoing Augustine in believing that it would be hard to convince rational men to follow a religion that denied things one could see for oneself.

Scrope's book was the culmination of extensive fieldwork in the Auvergne region, where dozens of conical hills made of loose piles of volcanic cinders overlook acres of black basalt. Deep valleys were carved into stacked lava flows on which these delicate cinder cones stood. Identical sequences of lava flows exposed in the walls on opposing sides of individual valleys proved that the river cut down into the lava. Lyell was intrigued by Scrope's description of how the lava flows buried the river gravels now exposed in the valley walls. Scrope's careful observations left no doubt that the lava had repeatedly filled a valley that the river just as often reexcavated. The layers exposed in the cliffs were not deformed and there was no evidence of catastrophic disruption. The valley-filling lava flows could be traced back to loose cinder cones sure to have been swept away by a flood capable of carving into hard rock.

The following May, Lyell set off to explore the region firsthand,

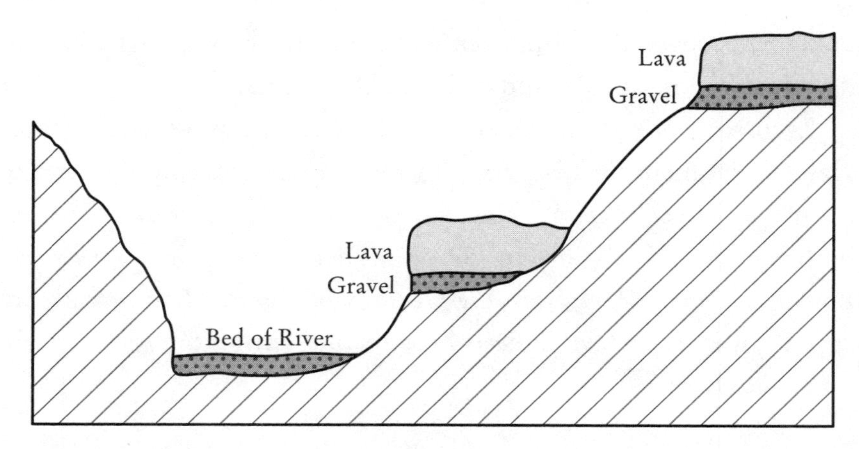

Lava flows emplaced over buried river gravels in Auvergne, France (based on Charles Lyell's 1833 Principles of Geology, *volume III, figure no. 61, p. 267).*

accompanying the influential Scottish geologist Roderick Murchison on an excursion through France. They visited Scrope's outcrops and studied the relationships between cinder cones, basalt flows, and river terraces. It quickly became clear to Lyell that a single flood could not have carved modern topography. Rivers slowly carved their own valleys.

From Auvergne, they traveled down the Rhone Valley to compare its rocks with those of the Paris Basin. Proceeding south into northern Italy, they traveled from Bologna to Florence and on to the Zoological Museum in Turin. Lyell realized that the rocks in the different parts of the regions they had just crossed had different fossils. Here was a formative realization for one who had never set out to become a geologist.

Fossils could be used to reliably assess the age of strata in southern Europe, something that could not be determined from mineral composition alone. The fossils in the younger rocks at the top of the regional geological pile were more like the modern fauna than were the fossils in the older rocks deeper in the section. The comings and goings of species from the fossil record could be used to track geologic time. Lyell was hooked. Here was the key to the grandest puzzle. The fossils in different rock formations could be read to tell geologic time. If you knew the mix of fossils in a rock formation, you could confidently deduce its age relative to other formations.

When Murchison returned to London in August, Lyell traveled on to Sicily, ending his career as a barrister. He was now a geologist, by accident rather than design. More than anything else his exploration of European geology convinced him of the enormous span of geologic time and that a global flood was not responsible for shaping the modern landscape. Perhaps Hutton was right after all. Maybe slow, steady change was the pace at which the world worked.

On his way back to England, in February 1829, Lyell stopped in Paris to compare the fossils he had picked up with those in the collections of French geologists. The proportion of still-living species increased farther to the south—and higher in the regional stack of rocks. Older rocks, lower down in the regional pile, held more species not represented in the modern fauna. This didn't square with the traditional biblically inspired view that, except for the Flood, everything's been the same since the Creation.

The trip through France and Italy convinced Lyell to try to sway public opinion away from the misconception that Genesis precluded the immensity of geologic time. It was an ambitious goal. Geological findings that contradicted conventional biblical interpretations weren't common knowledge, and geological audiences favored Cuvier's grand catastrophes to explain the geologic record. Few favored Hutton's style of uniformitarian thinking in which everyday processes slowly shaped the world. Writing for two audiences, Lyell tried to counter the dominance of catastrophist thinking among his colleagues without shocking the general public accustomed to the idea that Noah's Flood resurfaced our six-thousand-year-old planet. In 1830, he put his legal training to work in his *Principles of Geology*, building up an argument and defense against the reactionary outcry sure to follow.

In presenting his case, Lyell began with a history of geology that turned the uniformitarian-catastrophist debate into a simplistic choice. Things either happened catastrophically or they happened gradually. Casting the debate between uniformitarianism and catastrophism as between rationality and superstition, he decried the tendency of previous generations to conjure up grand catastrophes when the steady action of processes still operating today could explain the world.

Eager to make his mark challenging catastrophists, Lyell also was keenly aware of his own need to secure a steady income. Geologizing did not pay the bills. So with an eye on securing a chair in mineralogy or geology, and not wanting to be too provocative, he kept references to the Mosaic chronology and the biblical flood to a minimum.

Lyell staked out a position opposing the habit of invoking grand catastrophes to explain geological evidence.

> *We hear of sudden and violent revolution of the globe, of the instantaneous elevation of mountain chains, of paroxysms of volcanic energy. . . . We are also told of general catastrophes and a succession of deluges, of the alternation of periods of repose and disorder, of the refrigeration of the globe, of the sudden annihilation of whole races of animals and plants, and other hypotheses, in which we see the ancient spirit of speculation revived, and a desire manifested to cut, rather than patiently to untie, the Gordian knot.*[12]

In cataloging observations on physical processes now in operation, Lyell emphasized how erosion and uplift occur episodically. He calculated that it could take a big river like the Ganges more than seventeen centuries to carry away the tremendous mass of rock uplifted by a single great earthquake.

Lyell argued that the laws of nature governing geological processes remain constant, even though their effects vary through time.

Contemporary reviewers misinterpreted this as advocating no role for catastrophes in earth history. But this was not what Lyell meant. He described the tremendous erosive power of floods resulting from the failure of topographic barriers holding back lakes, specifically linking catastrophic floods with earthquakes and volcanic eruptions. With this nod to geological catastrophes, Lyell argued that processes still in operation could carry on for long enough to sculpt topography.

In dispensing with the need for divine intervention after the initial Creation, Lyell had taken one more step on the path toward full abandonment of a global flood as a geological reality. By the third volume of his *Principles* he explicitly dismissed the likelihood that a global flood ever happened. Any current capable of gouging deep valleys into hard rock would have swept away the fragile cinder cones of central France. Besides, Lyell's reading of Genesis implied a tranquil flood rather than Buckland's raging waters. That an olive tree remained standing demonstrated little, if any, scriptural support for erosion during the Deluge. He saw no case for a globe-wrecking flood.

Lyell suggested that a local flood could have wiped out the then inhabited world if there had been "extensive lakes elevated above the level of the ocean" in a region with "large tracts of dry land depressed below that level."[13] He went on to describe how this might occur in various places. An earthquake that breached the topographic barrier holding back Lake Superior would unleash a mighty flood down the Mississippi River valley. The low ground surrounding the Caspian Sea sat three hundred feet below the Black Sea. Breach the barrier between these inland seas and the lower basin would rapidly fill with rising water. Lyell speculated that if even deeper depressions had existed in the past, similar situations could have flooded what previously had been mountains. Here were plausible processes by which great floods might occur.

Despite his care to avoid confrontational language, the implications of Lyell's views were not lost on the panel reviewing him for appointment to a position at King's College in 1831, a post he desperately needed. The decision was in the hands of an archbishop, a pair of bishops, and two medical doctors, each of whom had the right to veto Lyell's nomination. When Lyell was informed of their concern about his unorthodox convictions, he fired off a letter to explain that although it was clear that the Flood could not have covered the entire planet, there was no evidence that "the whole inhabited earth . . . may not have been deluged within the last 3 or 4,000 years."[14]

Lyell's artful dance worked. He got the job and made a point of quoting one of the bishops to conclude his second lecture: "it is impossible that true religion can be injured by the ascertainment and establishment of any fact . . . [no science] affords a greater number of illustrations of the power & wisdom exhibited in the creation than Geology."[15] To Lyell, his geology demonstrated the manifest wisdom of the Creator, which meant the challenge lay in correctly interpreting both the rocks and the Bible.

Lyell's careful arguments and exposition mollified some, although by no means all, critics. Soon after Lyell's book was published, Sedgwick attacked Lyell's insistence on the uniform operation of processes through geologic time. Catastrophes were necessary to explain the deformation of strata and how ancient seabeds could be lifted up to form new land. Lyell's carefully constructed arguments may not have worked on Sedgwick, but they began to convert Buckland.

Within a decade, new discoveries convinced Buckland that Lyell was right. The volcanic cones of central France really were compelling evidence that valleys had not been incised by a global flood. Buckland's own fieldwork demonstrated that the drift, the great gravel sheet he had long attributed to Noah's Flood, was not deposited in a single event. There had been several episodes of deposition involving material from different sources. In his Bridgewater treatise

Buckland reveals the influence of Lyell's *Principles* when he states that the physical laws governing geological processes were as uniform as the law of gravity governing the orbits of planets.

It was Buckland who bore the brunt of clerical attacks after his abandonment of Noah's Flood. Conservative clergy may have seen Lyell as a godless radical, but they saw Buckland—the former champion of biblical geology—as a traitor. A new breed of scriptural geologists and clergy with limited knowledge of geological discoveries rose to defend Moses and attack Buckland. They recycled the discredited arguments of Burnet and Woodward and invoked Noah's Flood to explain secondary rocks, fossils, and the lay of the land.

In one of the least vitriolic clerical responses to Buckland's recantation, William Cockburn, Dean of York, claimed that there was no more to earth history than an initial six days of Creation and Noah's Flood about a thousand years later. A clergyman known for railing against what he saw as anti-Christian scientific ideas and theories, Cockburn revived even then discredited reasoning creationists still use to defend their preferred interpretation of Genesis. He ignored the work of Hutton, Cuvier, and Lyell.

Spelling out his ideas in a pamphlet attacking Buckland's new views, Cockburn attributed the formation of the primary rocks to the initial Creation after which primordial waters laid down the secondary rocks. Not much else happened until Noah's Flood, which therefore had to explain the entire fossil record. The bones of giant creatures lay in the oldest strata because these animals were too heavy for the ark and had drowned. Human remains were only found in unconsolidated surface layers and not in rocks because people fled to the highest peaks. There, they drowned some time after animals too confused to flee to higher ground had already become incorporated in flood-deposited sediments. In his rush to condemn Buckland for abandoning Noah's Flood, Cockburn simply dismissed the discoveries and evidence that had convinced the devout Buckland to abandon the idea of Noah's Flood as a geological event. In this way, Cockburn can be viewed as among the first modern creationists.

Several years later, in 1844, Cockburn had an ideal opportunity to challenge Buckland when the British Association for the Advancement of Science met in Cockburn's hometown of York. On the morning of the second day of the meeting, geologists flocked to witness the spectacle of Cockburn challenging their findings of the past forty years. With great composure, the stately Cockburn walked through the crowd and took the stage to stand by the society's president. In a brief presentation he laid out a theory purporting to explain all of geology as the result of a global flood. Cockburn insisted that the world's surface was shaped all at once. Geologists had to explain everything using Noah's Flood, including layered rocks. There had been no extinctions. Rivers did not cut their valleys. After Cockburn sat down and the raucous laughter had died off, Sedgwick rose to deliver a stinging hour-and-a-half response attacking Cockburn's woeful ignorance of geology in remarks characterized by an eyewitness as marked with "a scornful bitterness beyond the power of any reporter to reproduce."[16]

Cockburn was not easily silenced. Immediately after the meeting, he published his address as *The Bible Defended Against the British Association* and challenged Sedgwick to explain Earth's origin and evolution from the beginning to the present day. Opting not to answer at first, Sedgwick eventually wrote Cockburn a short note explaining that the antiquity of the world was demonstrated by unassailable geological evidence. Nothing if not persistent, Cockburn wrote Buckland and Murchison seeking to debate Earth's age. Neither was interested. Meanwhile, Sedgwick had written a long letter to Cockburn explaining his position and requesting the favor of no reply. Ignoring this collective dismissal, Cockburn decided that geologists were just afraid to debate. So he published his ideas as a

New System of Geology in 1849. That his book didn't catch on surprised few but Cockburn.

Buckland was not the only famous geologist to publicly reverse course on the flood. Less than a decade after Adam Sedgwick marshaled geological observations to show how a recent catastrophe reworked Earth's surface and deposited England's surficial gravels, he recanted, in his last presidential address to the Geological Society of London.

> *There is, I think, one great negative conclusion now incontestably established—that the vast masses of diluvial gravel, scattered almost over the surface of the earth, do not belong to one violent and transitory period. . . . We had, in our sacred histories, the record of a general deluge. On this double testimony it was, that we gave a unity to a vast succession of phænomena, not one of which we perfectly comprehended, and under the name diluvium, classed them all together. . . .*
>
> *Our errors were, however, natural, and of the same kind which led many excellent observers of a former century to refer all the secondary formations of geology to the Noachian deluge. Having been myself a believer, and, to the best of my power, a propagator of what I now regard as a philosophic heresy, and having more than once been quoted for opinions I do not now maintain, I think it right, as one of my last acts before I quit this Chair, thus publicly to read my recantation.*[17]

With this spirited reversal, Sedgwick joined Lyell in arguing for disentangling geology from the biblical flood. It was becoming apparent that the stories in Genesis were too short and mysterious to either confirm or challenge geological theories.

In the 1830s the question was not when Noah's Flood occurred but how many grand catastrophes the world had seen. Agreement was growing that there was more to Earth's story than just what the

Bible said. Moses did not lay it all out. Many worlds had come and gone since the dawn of time. Shortly after Buckland's recantation, the Swiss naturalist Louis Agassiz explained the surficial debris and stray boulders of northern Europe. The evidence traditionally interpreted as resulting from a global flood actually recorded the action of glaciers that overran Europe during an age of ice, leaving Noah out in the cold.

By the 1850s, Christian men of science overwhelmingly believed Earth was extremely old. In the decades before Darwin, the failure of a literal interpretation of Genesis to account for earth history helped create new rifts in Christian philosophy. In the spirit of Augustine, many Christians adopted the view that geology could help guide reinterpreting biblical stories. Others, without a background in natural philosophy or geology, came to be known as scriptural geologists. They either considered a literal interpretation of the Bible paramount and geology mistaken or embraced the idea that God just made the world look old, hiding fossils in rocks back at the initial Creation. In this split lay the roots of modern creationism.

Cockburn may have failed to convince the British Association, but he was by no means a lone voice. Scriptural geologists with little to no geological training ignored problematic geological evidence, promoted discredited theories, and invoked exceptions to biblical literalism when it suited their arguments. These forerunners of modern creationists banded together against the coalescing views of ever more geologists who rejected the idea that the Creation and Noah's Flood were all there was to earth history.

Today geologists view all processes as fair game—from slow and steady everyday change to dramatic catastrophes. It's not one or the other, as Lyell and Cockburn both portrayed things. Over the past several centuries, generations of geologists built their ideas on top of preceding theories, disproving or reinforcing what they had heard before. In the process, they learned how everyday change really does add up to big effects—given time—and that geological catastrophes really did happen, causing mass extinctions not just once but at least five times in the history of the world.

Along the way, the tension over how to read the geologic record—whether as an unimaginably long progression of everyday events or as a series of grand disasters—has characterized the earth sciences. Misunderstanding the nature of this tension caused friction in the relationship between geology and Christianity and still fuels conflict between science and religion.

By the end of the nineteenth century, geologists had disproved a young Earth and a global flood. Archaeologists, however, had begun to unearth ancient flood deposits in the sandy floodplains of Mesopotamia, setting off new arguments for and against evidence thought to record the biblical flood. Their discoveries carried startling implications about the age and origins of the biblical flood story.

8

Fragmented Stories

Squinting in the dim light of a windowless, unheated basement room of the British Museum, George Smith rose slowly from his seat stunned by what he'd just read. Spread out before him in neatly reassembled baked clay fragments lay the story of Noah's Flood—or at least the basic elements of it. The blocky symbols of ancient cuneiform told of a divine warning about an impending flood conveyed to a righteous man, the building of a great boat, the riding out of days and nights of rain, and the eventual stranding of the boat on a mountain when the floodwaters receded. Smith's excitement echoed throughout the museum. How could the biblical flood story be inscribed on a broken clay tablet excavated from a Sumerian library older than the Bible itself?

It was a shocking revelation. Who in Victorian England or among Christians around the world would have imagined that the story of the biblical flood was a degraded pagan myth and not the other way around? And yet, Smith had just uncovered tangible proof that the biblical flood was a recycled Babylonian story.

Running around the room in exhilarated agitation, Smith shed his jacket and tie, shocking co-workers attracted to the commotion. Normally such behavior might have gotten him fired. But his puzzled colleagues tolerated his odd demeanor as word quickly spread about the assistant curator's astounding discovery.

Born in 1840, Smith became obsessed early with Mesopotamian archaeology. He eventually entered an apprenticeship with a banknote engraver, though he was far more drawn to fascinating accounts of excavated Assyrian palaces. Intrigued with explorer Henry Rawlinson's discovery of how to translate the cuneiform alphabet, Smith dreamed of resurrecting the stories preserved in the columns of tiny wedgelike characters impressed into clay tablets. He spent his meager income on obscure textbooks and his evenings learning to read arcane inscriptions and mastering a dead language. After work he haunted the British Museum, where the staff noticed

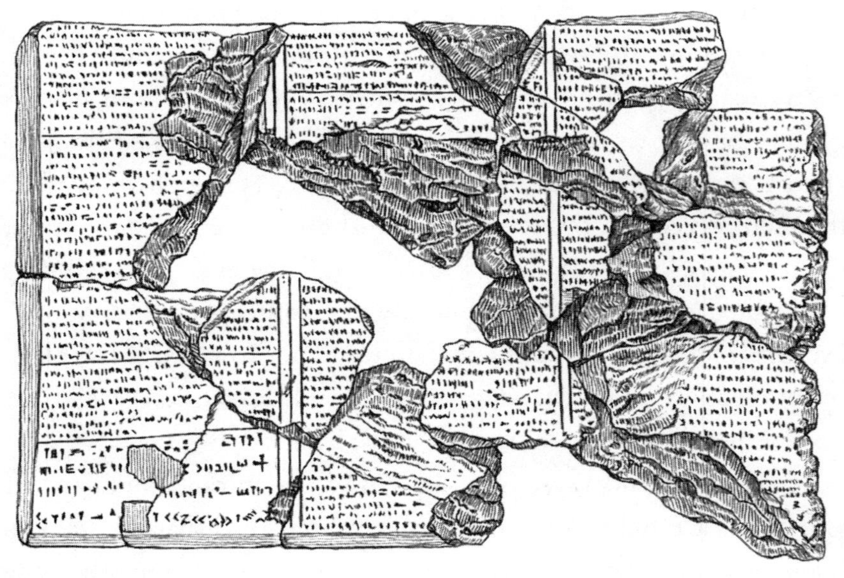

George Smith's reconstructed cuneiform tablet of the Babylonian flood story (by Alan Witschonke based on an illustration in Smith, G., 1876, The Chaldean Account of Genesis, *Scribner, Armstrong & Co., New York, p. 10).*

the enthralled youth's interest in the collection of fragmented clay tablets. Who knew what mysteries lay hidden in the thousands of fragments in the museum's collection?

For half a decade, from 1849 to 1854, archaeological expeditions returned crates containing thousands of clay tablets to the British Museum. Digging through the rubble of ancient Nineveh, near the modern Iraqi town of Mosul, excavators discovered the ruins of King Ashurbanipal's library dating from around 670 BC. Not recognizing their significance at first, the museum's curators thought the tablets were decorated pottery. After minimal precautions were taken to protect them on the way to London, crates full of broken tablets arrived at the museum and sat neglected in storerooms.

All those worthless fragments turned out to be the remains of the world's oldest books. The secrets of a dead civilization lay scattered in countless pieces of an archaeological jigsaw puzzle. Smith's knowledge of cuneiform uniquely qualified him for the job of sorting fragments excavated from the rubble of the Royal Assyrian library. The museum hired him in 1863 as a curator's assistant.

He faced quite a challenge. Some tablets were broken into more than a hundred pieces. Reconstructing them would be a tedious task, ideal for a detail-oriented introvert. Smith threw himself into his job and was soon matching tiny pieces of broken clay together. A natural at grouping fragments by color and shape, he had a remarkable knack for reassembling the jumbled pieces into whole tablets.

For almost a decade the quiet curator's assistant painstakingly pieced tablets back together, patiently working through the museum's collection. Then, one damp fall morning in 1872, he noticed references to the creation of the world. He soon found a large fragment on which two of the original six columns of writing were intact, two were half-preserved, and two were missing. It seemed to tell of a great flood.

But only part of the intact fragment was legible; the rest lay covered beneath a thick white deposit. Frustratingly, the curator in charge of cleaning tablets was away, and Smith was not authorized to take on the task. Naturally high-strung and nervous, Smith became increasingly agitated waiting for the curator to return. When he finally did, Smith pounced on the cleaned fragment.

Scanning down the third column, he struck gold.

> *My eye caught the statement that the ship rested on the mountains of Nizir, followed by the account of the sending forth of the dove, and its finding no resting place and returning. I saw at once that I had here discovered a portion at least of the Chaldean account of the Deluge.*[1]

The partial account Smith described was a speech given by a character he provisionally named Izdubar (who eventually came to be known as Gilgamesh after scholars refined their understanding of Sumerian). Recalling Izdubar's name from other fragments, Smith searched for them and gradually reconstructed the tablet, piecing the story together as he completed the second column. He then found and reassembled additional, overlapping copies that filled in the sixth column and nearly completed the first column. It was like multiple editions of the same book. Further investigative work turned up more fragments, nearly completing an account of a great flood.

Surprisingly, the story paralleled the biblical story. The mighty King Izdubar had conquered monsters and united the feuding kingdoms between the Tigres and Euphrates but fell ill in old age. Fearing man's last enemy, death, he sought out Sisit (later translated as Utnapishtim), the immortal survivor of the great flood the gods sent to destroy humanity. Warned of an impending flood, Sisit built a ship and caulked it with bitumen before loading his family and animals aboard to ride it out. After seven days and nights they ran aground on a mountainside and Sisit sent out a dove, a swallow, and finally a raven to search for dry ground.

While this ancient cuneiform narrative was similar to the more recent biblical story, Smith saw more differences between the two stories than just the number of days and nights of rain (seven versus forty). The Mesopotamian story alluded to a maritime tradition. The ark was called a ship. It had a pilot wise enough to take it on a trial voyage before the flood arrived. In contrast, the biblical story suggested inland authors unfamiliar with seafaring. The biblical ark was simply described as a great box. Did the Babylonian and Hebrew stories represent different versions of the same events? Or was the biblical flood a reworking of the Babylonian story?

On December 3, 1872, Smith presented his findings to the Society of Biblical Archaeology, sharing the stage with the prime minister and the dean of Westminster. His lecture captivated scholars and the general public alike. Newspapers trumpeted the discovery

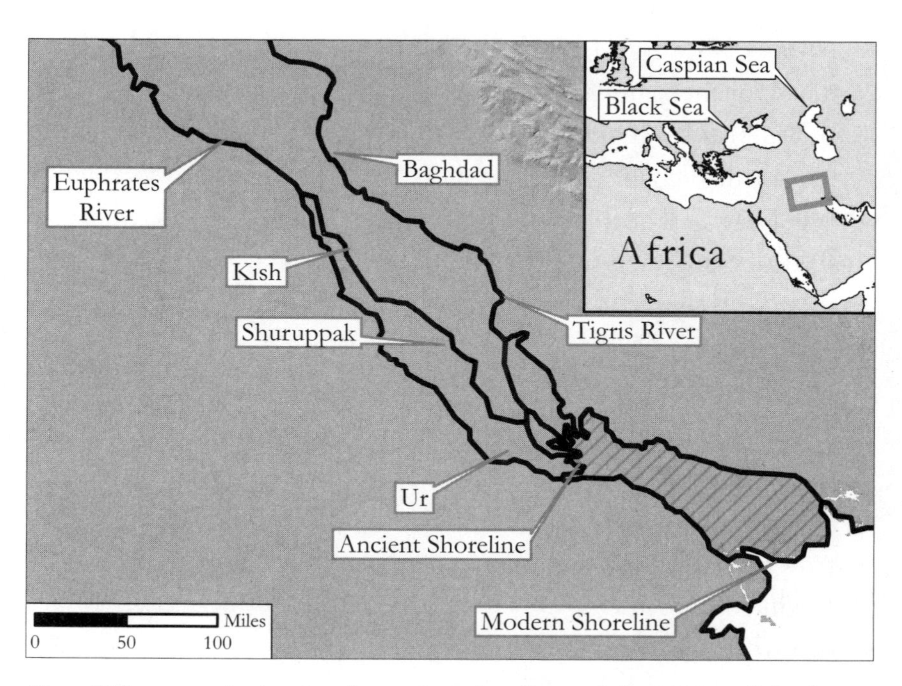

Map of Mesopotamia showing the modern shoreline and the position of the shoreline in ancient Sumeria when Ur and Shuruppak were in the coastal estuary.

of a prebiblical source for the biblical flood story. Immediately after his presentation, the *Daily Telegraph* offered Smith the princely sum of a thousand guineas to search for more tablets at Ashurbanipal's ruined library. The British Museum jumped at this publicity bonanza, granting Smith six months' leave.

With no field archaeology training, and after digging through the ruins for just eight days in May 1873, Smith found a fragment that completed the first column of the tablet under reconstruction at the British Museum. It filled in the part of the story that included the command to build a ship and load it with animals. Near the end of his trip, Smith also found fragments of additional tablets describing the creation of the world in six days as well as man's temptation and fall.

In unearthing multiple copies of the same stories, Smith discovered how the Genesis stories grew out of much older texts. The Assyrian king was apparently a bibliophile whose agents sought out inscribed tablets for his literary treasure house. Multiple tablets with different versions reflected the evolution of the flood story. Some versions dated from long before Ashurbanipal's rule. It could not be considered coincidence; Smith kept finding more and more evidence corroborating a prebiblical flood story.

Smith thought more than ten thousand inscribed tablets were originally housed in the upper floors of the ruined palace. Apparently arranged by subject, some tablets formed a series, the longest of which consisted of over a hundred individual tablets. Each shared the title that began its series, and each was numbered with its position in the series and started with the last phrase of the preceding tablet.

This once well-organized library lay in ruins. Scorch marks showed that many tablets broke apart during the fiery destruction of Nineveh. Subsequent treasure seekers also took a toll, tossing tablets aside in the quest for better loot. Finally, cycles of rain and drying splintered most tablets into piles of clay shards.

Smith shipped crates and crates of fragments back to London. As he fitted them back together he discovered that the flood story was the eleventh of a twelve-tablet series. Different tablets revealed several distinct versions. One nearly complete tablet revealed that the gods sent a great flood to destroy the city of Shuruppak. This version referred to the flood survivor as Atrahasis, who, like Sisit, built a ship, sealed it with bitumen, and loaded it with his wealth, family, and beasts of the field. As in the other version, the great flood raged over the surface of the earth for seven days and nights, killing all living things. After the ship came to ground on a mountain, Atrahasis sent out a dove, then a swallow, and finally a raven before disembarking after the waters receded.

Henry Rawlinson, Smith's mentor who, decades before, stumbled onto the key to deciphering cuneiform, seized upon the twelve tablets as proof that the flood story was a solar myth tied to zodiac symbols in which each tablet corresponded to a different sign. The tablet that contained the flood story corresponded to the eleventh month, the rainiest time of the year, the month ruled by the storm god.

But Smith thought this ancient story from Ashurbanipal's library recorded an ancient catastrophe dating back long before the Bible. Maybe the Jews adapted an older Babylonian story to monotheism. Smith composed a table showing how basic elements occurred in the same order in the biblical and Babylonian narratives. However, he saw enough differences in the details to believe the stories represented distinct traditions recording the same events. Perhaps the mountaintop on which the ark landed was a Mesopotamian temple, rising above the floodwaters and offering a beacon of hope to anyone adrift on the submerged lowlands.

After returning to England from his second expedition in 1874, Smith focused on combing through the thousands of fragmented tablets to reconstruct those that told the history of the world from the creation to the flood. He found tales of the building of the Tower of Babel and of the Confusion of Tongues. In their account of the world's creation, the cuneiform tablets told of the initial chaos from which the universe was made and how, after each step along the way, the gods pronounced their creation good. Smith even found a tablet telling of the fall of a celestial being corresponding to Satan.

His luck eventually ran out on his third expedition. He ignored the advice of locals at his dig and set off for Syria during the hottest part of the summer. After contracting dysentery, he died along the way, in August 1876.

Smith's astounding discovery upended conventional thinking about the origin of flood stories. His conclusion was revolutionary: key parts of the Old Testament were adapted from older pagan tales. Until then, Christians generally argued that pagan flood stories from other cultures were rooted in the biblical story. After Smith's revelation, even conservative theologians began to concede that the story of Noah's Flood lay rooted in an historical Mesopotamian flood rather than a global disaster.

Smith's startling proof that the biblical account of the Flood originated in older Babylonian stories set off a scramble among archaeologists to find Mesopotamian flood deposits. Everyone believed that evidence for a civilization-ending flood could be found there. This soon became a nagging problem, as archaeologists were not able to find evidence for such an enormous flood and fell into arguing over which of their local flood deposits recorded the biblical flood. Like geologists in the eighteenth and nineteenth centuries, twentieth-century archaeologists faithfully searched for evidence of the Flood.

In 1922 British archaeologist Leonard Woolley began excavations at the biblical patriarch Abraham's hometown, the ancient city of Ur, along the Euphrates River near the modern town of Nasiriya in southern Iraq. Convinced only a combination of unusual circumstances could turn typical delta flooding into the biblical flood, Woolley dug for evidence of a catastrophic flood. He eventually found what he was looking for in more than ten feet of well-sorted, water-laid silt that buried a ruined city. Three additional feet below layers of ash, rubble, and pottery fragments lay the soil upon which southern Mesopotamia's earliest farmers had built Ur. Long before Abraham's day, an ancient flood had buried the birthplace of the biblical patriarch.

When he found a similar sequence of flood deposits burying cultural debris at two more locations near Ur, Woolley claimed to have unearthed deposits from a great flood that swept away early villages. He lost no time telegraphing London to report his supposed geological footprint of the biblical flood. Returning the following year, Woolley's team found ten feet of water-laid sand deposited atop yet more cultural debris at another location. Convinced he had found evidence of a regional flood, he concluded that here, surely, was the signature of Noah's Flood.

Woolley's discovery was a sensation. The news he had uncovered evidence of the biblical flood electrified the public as it spread across headlines, radio, and newsreels. Suddenly, the hunt was on again to find more proof of Noah's Flood.

Working at Kish, an ancient Sumerian city well upstream of Ur and eight miles east of Babylon, a team of Oxford archaeologists led by Stephen Langdon found more flood deposits. Langdon's and Woolley's teams promptly began bickering about who had unearthed the biblical flood. Defending the sanctity of his deposit, Woolley maintained that eight layers of sediment containing distinctively different cultural debris, and therefore representing the coming and going of several societies, separated the Kish and Ur flood sands. Woolley insisted that Langdon's deposits could not represent the same flood. Naturally, his Ur flood was the real Flood; Langdon's later Kish flood, Woolley maintained, was just another garden-variety Mesopotamian flood.

Soon both Woolley's and Langdon's stories were called into

doubt by archaeologists' inability to find similar deposits at nearby Tell Obd. Subsequent borings and trenches revealed Woolley's flood deposit could not be traced very far. All signs pointed to a local deposit formed when a burst levee inundated a few square miles of floodplain. If one of these deposits recorded Noah's Flood, it was a very local affair.

Through decades of academic squabbling, Woolley promoted his Ur flood as the real thing. In 1956, writing in the *Palestine Exploration Quarterly*, he claimed that cuneiform tablets dividing the reigns of Mesopotamian kings into periods before and after the Flood confirmed his discovery. Entombed beneath the silt at Ur lay ruined houses with distinctive pottery characteristic of the earliest settlements. Above the lowest layer containing cultural debris, the pottery changed to a different style that he interpreted as belonging to a new culture that arrived from the north. Woolley believed his Ur flood destroyed everything in the delta except the largest towns, which had grown tall enough to rise like peaks above the floodwaters.

From everything he'd seen, Woolley concluded that the story of this flood was part of Abraham's cultural heritage from Ur. The district of Haran, where Abraham subsequently lived, even had a version of the flood story in which the name of the hero was similar to "Noah." Woolley argued that Abraham's family had adopted the local flood story, purged it of all references to false gods, and handed it down through oral tradition to become the basis for the story recorded in Genesis.

In 1964, British archaeologist Max Mallowan, the husband of mystery writer Agatha Christie, summarized the evidence for a prebiblical Mesopotamian origin for the story of Noah's Flood. Mallowan considered the biblical story to have come from an oral account of traumatized survivors of a regional flood. Sumerian scribes subsequently preserved the story on clay tablets of the type George Smith would eventually reassemble and translate. But none of the flood deposits that archaeologists were squabbling over had been large enough to belong to a flood capable of wiping out all of Mesopotamian civilization. If one account of flooding was the source of the biblical story, it was the tale of a local disaster that developed into the myth of a global flood.

Although there was no consensus among archaeologists as to which, if any, of these deposits was from Noah's Flood, when the Tigris River flooded in 1954 and submerged the floodplain for hundreds of miles around Baghdad, it alerted everyone to the reality that enormous floods could submerge the area. Surely, some thought, such events could have been recorded in Mesopotamian flood stories. Despite bitter arguments, archaeologists generally favored the idea that the origin of the strikingly similar Sumerian, Babylonian, and biblical stories lay in catastrophic flooding along the Tigris and Euphrates rivers. This made sense; after all, to the residents of Mesopotamia, their home was the entire civilized world.

It's nearly impossible today to understand how gargantuan ancient floods were, because today so many of the world's rivers have been engineered to reduce floods. To imagine the devastating effects of an unusually large flood on an ancient low-lying region, we can look at the 2008 flooding of Burma's heavily populated Irrawaddy River delta, where in some areas nine out of ten inhabitants drowned overnight. The populated lowlands filled up like bathtubs when the levees broke. The story of a great flood that submerged the world would have been perfectly plausible to those living in Mesopotamia's flood-prone estuary, where everyone was no more than a few generations removed from a locally disastrous flood.

By the time Smith took his ill-fated trip to Syria, he realized that the ancient tablets that so captivated him recorded multiple versions of the story of a great flood. As it turned out, Smith discovered portions of at least three flood stories that predated the biblical story by centuries, if not millennia. The earliest, a Sumerian version, featured Ziusudra as the hero. The middle version, the Akkadian story of Atrahasis, was later integrated into the third version, the Gilgamesh epic, with Utnapishtim (Sisit) as the Babylonian flood survivor. Smith's discoveries showed that Mesopotamian flood stories had a long and complex history dating back to the frontier between mythology and history.

The earliest version of the flood stories that Smith uncovered preserved an older tale inscribed around 1600 BC. This Sumerian version of the story told of the flooding of Shurrupak, a city about 30 kilometers north of Uruk in southern Iraq. Another version divides history into the time before and after the flood and names Ziusudra as the last pre-flood king of Shuruppak. Excavations at Shurrupak revealed that a flood did indeed destroy the city around 2800 BC. Perhaps the story of a flood that destroyed the city circulated for a thousand years before it was pressed into clay and baked for posterity.

The surviving fragments of the Sumerian version open with a speech by the supreme god Enlil telling how he established kings to rule over each of the five Sumerian city-states. When the capricious gods later decided to destroy mankind, pious Ziusudra overheard from a sympathetic god that a great flood was coming. So he built a large vessel and rode out the flood for seven days and nights. After making appropriate offerings to the gods, he was rewarded with eternal life for having saved humanity.

This even-then ancient story served the political establishment of Mesopotamia by reinforcing the divine sanction of kingship and promoting the interests of priests who kept the temples. Whatever its origin, the Sumerian flood story proved useful enough to the ruling class that when King Hammurabi conquered Sumer and founded the Babylonian empire around 1800 BC, the narrative was rewritten and characters renamed in Akkadian, the language of Babylon.

The earliest copy of the middle version of the flood story (starring the hero-king Atrahasis) dates from around 1635 BC—a little before the earliest surviving copy of the much older Sumerian story was created.

The Akkadian version begins with the lesser gods toiling in the fields to maintain the all-important irrigation system used to grow food for the greater gods. After decades of backbreaking work, the lesser gods rose up, burned their tools and stormed the chief god Enlil's house. Roused from sleep, Enlil called an assembly and sought the advice of Enki, god of fresh waters, who proposed solving the dilemma by creating people to work the fields.

This worked well for a while, but after 1,200 years people had been so fruitful and had multiplied so prolifically that the constant commotion of human society disturbed the gods. Annoyed at being kept awake, cranky old Enlil sent a plague to quiet the land. After another 1,200 years, the problem recurred. So Enlil sent a great drought. But again, after another 1,200 years, noisy carousing kept Enlil up at night. Withholding the field-watering annual flood bought another millennium of peace and quiet. Then, when the infernal racket began all over again, Enlil had truly had enough. This time he planned to send a great flood to destroy humanity for good.

Each time that the angry god sought to exterminate the human pests he regretted releasing upon the land, Enki had thwarted his superior's genocidal plan by tipping off the mortal King Atrahasis in time for some people to survive. Enlil finally realized that a lesser god was leaking his plans, so he swore them all to secrecy about the coming flood. This time, Enki loudly told the plan to the wall of Atrahasis's reed hut. Atrahasis overheard the warning and converted his home into a boat, which he loaded with his family, possessions, animals, birds, and grains—everything he would need to re-create human society after the flood.

The makeshift boat rode out the storm for seven days and seven nights and then ran aground on a mountainside. After another seven days passed, Atrahasis sent out a dove to seek land. The dove returned unsuccessful. Atrahasis then sent out a swallow, also unsuccessfully. Finally, with the waters receding, he sent out a raven, which, finding land, did not return. The story ends with Atrahasis disembarking and sacrificing a sheep and burning incense offerings to the gods.

While the original Sumerian story shares striking details with that of Noah's Flood, the parallels to the biblical story are even more apparent in the later elaborately detailed Babylonian flood story of Gilgamesh. Fearing death, Gilgamesh sought the secret of eternal life from Utnapishtim, the great king who saved mankind from the flood. Passages that are virtually identical show that the tale of Atrahasis was spliced into the Gilgamesh epic, with the name of the heroic flood savior changed to Utnapishtim (which some consider an old Babylonian translation of "Ziusudra"). One version of the Gilgamesh epic even refers to Utnapishtim as "Atrahasis."

As Smith and others continued to find and translate more versions of the flood story, its historical background grew increasingly complicated. Each period and region possessed its own version, with no master version against which to compare all other versions. There were many versions of the Mesopotamian flood story. Societies throughout the region adopted the tale, adapting it to their language and culture.

The story of a great flood became widely known across the Middle East because Akkadian, the language of Babylon, served as the language of diplomacy until the first millennium BC. Novice scribes helped spread the story from one culture to another as they practiced their Akkadian by copying classic texts. It has even been argued that an abbreviation of Utnapishtim, with emphasis on its second syllable, was pronounced as "Noah" in early Palestine. As a foundational piece of regional lore, it's a story the Jews would have been exposed to as they wept in captivity by the rivers of Babylon after their exile from Judea.

On the whole, the exile of the Jews to Babylon was a period of political banishment rather than outright enslavement. The Jews were treated well enough in their temporary home so that significant numbers chose not to return to Judea when their captivity ended. We know that at least some Jews rose in Babylonian society, if only because the Bible says that those who returned to the Holy Land dragged their own slaves with them. That they also took the Mesopotamian flood story fits the expected pattern in which a well-treated conquered people are more likely to assimilate their captor's culture.

Still, the Genesis stories differ from Babylonian precursors in a very fundamental way. The contrasts between monotheistic and polytheistic culture is striking, and reading Genesis as literature intended to promote monotheism is illuminating. Genesis lists the pantheistic gods and says that one true God created them all. It is an epic poem with a purpose. Earth, sky, sun, moon, plants, and animals—they are not gods. According to Genesis, sea monsters were created on the fifth day.[2] This explicitly refutes the Mesopotamian creation story in which the patron god of Babylon subdued the forces of chaos, slaying the angry goddess that ruled the cosmic sea to create the world and everything in it. Here, perhaps, we find the original aim of the opening chapters of the Hebrew Bible: refuting the account of Creation posed by the polytheistic Mesopotamian culture.

The Babylonian flood story was even known to the ancient Greeks. The obscure historian Alexander Polyhistor attributed an account of a great flood to the Babylonian priest Berossus, who lived in the time of Alexander the Great several hundred years after the Jews were exiled to Babylon. Writing in the first century BC, Polyhistor recounted how the god Kronos ordered Xisuthros (likely a phonetic transliteration of Ziusudra, the original Sumerian flood hero) to build a boat that could carry his family and friends through a flood sent to destroy the rest of mankind. He was to stow provisions, animals, and birds on board and then sail off as the flood rose. Later, as the flood receded, Xisuthros set the birds free, only to have them return, unable to find land to rest on. The second time he sent them out, the birds returned with muddy feet. The third time they did not return at all. Finally, the boat ran aground. There, on a mountain, Xisuthros built an altar and offered a sacrifice to the gods for delivering him through the ordeal. The similarities between Polyhistor's story and both the Sumerian and biblical flood stories are clear.

The Greeks also had a flood story, although theirs evolved to parallel the Old Testament story. In the ninth Olympian ode of Pindar, Deucalion and his wife, Pyrrha, come down from Mount Parnassus (the highest point in southern Greece) to populate the world after Zeus drained the floodwaters, revealing the fertile lowlands where humanity might thrive. In the fourth century BC, Plato taught that Deucalion's flood was a local affair that only covered the plains, allowing those who fled to the hills to save themselves. In both traditions, Deucalion and Pyrrha were the ancestors of all Greeks.

The best-known version of Deucalion's story is found in Ovid's *Metamorphoses*, a Roman elaboration of Greek myths in which Prometheus warned righteous Deucalion and Pyrrha of a great flood the gods were sending to punish mankind. The pious pair built a ship, loaded it with provisions, and rode the rising floodwaters to the only unsubmerged peak. As the flood receded, they disembarked on Mount Parnassus and thanked the gods for delivering them from the deluge. Finding themselves alone in a devastated world, they went to Themis's shrine to ask how mankind might be restored. Advised to throw the bones of the great mother (Earth) behind them, the lonely couple tossed stones over their shoulders. Deucalion's stones became men and Pyrrha's became women.

The satirist Lucian's second-century retelling of the story expanded to include a great ark onto which Deucalion loaded pairs of every kind of creature on Earth. The Greek flood story was evolving to track the increasingly popular Old Testament story. Whether or not they originally represented different versions of the same event, ancient flood stories were transmitted from one culture to the next, demonstrating the attraction this story held for succeeding Middle Eastern societies.

Thousands of miles to the east, on the far side of Mesopotamia, Hindu society also had flood traditions. In the earliest version, recorded in the *Satapatha Brahmana* sometime between the fourth and second centuries BC, a tiny fish swam into the hands of a man named Manu as he was washing himself. The fish called out, "Rear me, and I will save you." When Manu asked what it would save him from, the fish replied that one day a great flood would carry away everything. So Manu raised the fish in a jar, and then a pit, until the fish grew large enough to avoid predators. He then returned it to the sea. The grateful fish told Manu when to build a boat, and as the flood came the now very large fish towed Manu to a Himalayan peak and helped him fasten his boat to a tree. When the floodwaters receded, Manu found himself alone in the world and began to pray. His prayers were answered within a year when a woman grew from his offerings of butter and sour milk. The new couple enthusiastically set about repopulating the world.

Later iterations of this story demonstrate its evolution, but could it originally have come from the Babylonian flood story? Possibly. Indian seals and jewelry found in Mesopotamian excavations document exchange between the two cultures as early as 2500 BC. Sea trade routes provided for cultural exchange in later times. Such links led some to argue that the Mesopotamian flood story spawned Hindu flood stories. Advocates of this view point to the basic plot as paralleling the broad outline of the biblical flood.

There are striking differences, however. Foremost among these is that in Hindu cosmology a great flood ends each era of the world, repeatedly wiping out humanity. Unlike Noah's Flood, Manu's flood is not a unique event. It was just one of many world-destroying floods. Other Indian flood myths variously invoked a rain of fire or food shortages that tempted people to desecrate sacred trees holding the proverbial forbidden fruit. These causes are quite different from those in the Hebraic tradition, in which debauchery and wickedness become the root causes of the flood, and the Mesopotamian tradition, in which humankind is destroyed for being a general nuisance. Perhaps these differences reflect local embellishments as the flood story traveled beyond Mesopotamia.

While the origins of these differences are unknown, what is certain is that flood stories evolved over centuries in the retelling, regardless of how or why they originated.

As geologists abandoned Noah's Flood as an explanation for the world's topography and archaeologists kept digging for Mesopotamian flood deposits, literary scholars professionalized the study of the history of the Bible itself. Paralleling the emergence of geology as a secular profession, historians began to formalize Bible studies, approaching the study of the Old Testament with the same independence and intensity geologists used to study rocks. Traditional interpretations of the Bible faced new trials as literary scholars concluded that Genesis was compiled from older sources.

9

Recycled Tales

Centuries before George Smith discovered that the opening chapters of the Bible were reworked Babylonian tales, controversy over the authorship of the Bible centered on how to interpret it as the literal word of God. The original Hebrew version had no vowels, leaving room for interpretation as to the specific wording when the Bible was translated into Greek. In 1538, the Jewish scholar Elijah ben Asher Levita demonstrated that the accents and points indicating where to insert vowels, add punctuation, and divide Hebrew words were invented by rabbis long after the translation of the Jewish Bible. Before the adoption of modern written Hebrew, the Jewish Bible consisted of a string of consonants. Meanings could vary depending upon how one inserted the missing vowels and where one divided words. Biblical translators like Saint Jerome had to use their judgment, which could introduce varying shades of meaning and complicate literal interpretations. Concern over potential human influence and errors came to a head in 1650, when Louis Cappel, a French Calvinist professor of biblical studies, painstakingly compared biblical translations in his massive *Critica Sacra* (*Sacred Criticism*) to demonstrate that the Bible was a book with a history, rather than the word of God delivered directly from the source.

Even before the Renaissance, it was well known that there were striking differences between the Greek and Hebrew Bibles. Arguments over which Bible was the true word of God led some scholars to argue that the Hebrew text was corrupt, or had even been intentionally altered to deceive Christians. Others argued that the Greek Bible was a hodgepodge of inferior translations, or that the Latin Bible was full of errors. This presented Christians with the awkward challenge of which version to believe.

Grappling with this controversy, Martin Luther emphatically labeled the Latin Bible a flawed text and devoted himself to sorting through the different wordings of the Greek and Hebrew texts. His efforts led him to dismiss the books of James and Jude as true scripture and relegate them to the end of his Bible. About the Book of Revelation, he wrote that every man could make up his own mind, but he "cannot find that it was inspired by the Holy Spirit."[1] In his view, the Latin Bible was so filled with errors that to confidently discern the meaning of scripture, one had to go back and read the original texts. Even Luther acknowledged that scriptural interpretation required care to avoid potentially flawed plain-sense understanding.

When the sixteenth-century Council of Trent met to judge Protestant heresies and clarify church teachings, the assembled bishops were deeply concerned that if they upheld Luther's critique the Latin Bible would lose all authority. The more they debated, the more authoritative Jerome's Latin translation became. The council finally declared the Latin Bible superior to the Greek and Hebrew versions, a conclusion they considered to be inspired by God. The bishops disagreed with Luther's claim that ordinary men could interpret the plain words of scripture for themselves. Fearing that freedom to interpret the Bible for oneself was the first step on the road to heresy, the Council moved to protect the church's authority and, in a fit of brinkmanship, deemed the translation more authentic than the original.

Recognition that Moses did not write much of what was attributed to him caused quite a scandal in 1685 when French clergyman Richard Simon outraged both Protestants and Catholics alike with his *Critical History of the Old Testament*. He came to this conclusion when his superiors in the Catholic Order of the Oratory asked him to provide scholarly arguments for use against Calvinists who rejected the authority of the church and trusted the Bible alone for spiritual guidance. Critically dissecting the Bible, he turned his attention first to the opening chapters of the Old Testament. The contradictions and confusion attending various literal interpretations of scripture could be explained by recognizing the historical nature of Genesis as a compiled story. The conflicting styles, repetitions, and logical impossibility of Moses writing about his own death implied that Genesis was compiled by a series of scribes long after Moses died. Extending his analysis to the New Testament, Simon was able to demonstrate that no original version of the Bible survived; that variations and contradictions had crept into the text as vowels, words, and whole passages were lost, added, or modified over centuries of translation and transcription.

His attack on biblical inerrancy—the belief that the Bible held no errors whatsoever—shocked both the Calvinists it was supposed to shock and the Catholic Church that commissioned the work. Simon believed scripture to be divinely inspired. He just did not know which of the modern versions corresponded to the original one. As a reward for a job done too well, his book was banned and he was expelled from the Oratorians.

Half a century later, the censors of the Sorbonne ignored French physician Jean Astruc when he advanced the same argument. Noting the striking repetition of events in the biblical flood story and the use of two names for God, Astruc claimed that Moses compiled Genesis from even then ancient accounts handed down from the patriarchs. Astruc's suspicions were based on several lines of evidence. First there were the unnecessary repetitions, like the two creation stories of Genesis 1 and 2. Then there was the story's awkward jumping back and forth through time. Astruc saw these anomalies as originating when Moses merged several original versions into a single story. The Bible was starting to be seen as a book that had evolved.

In the late 1700s, German intellectuals introduced more formal literary scholarship into biblical criticism. Johann Eichhorn, the prominent professor of Oriental languages at Jena University, compared biblical narratives and concluded that many of the stories in Genesis were fanciful accounts of prehistoric events. Analyzing the style of different passages, he sorted through and disentangled a literary stratigraphy that revealed Genesis to be a composite story.

In revolutionary America, where conventional institutions were no longer sacred, Thomas Paine took up the implications of Eichhorn's conclusion to attack the Bible in the name of the Enlightenment in his pamphlet *The Age of Reason.*

> *Take away from Genesis the belief that Moses was the author, on which only the strange belief that it is the word of God has stood, and there remains nothing of Genesis, but an anonymous book of stories, fables and traditionary or invented absurdities, or of downright lies. The story of Eve and the serpent, and of Noah and his ark, drops to a level with the Arabian Tales, without the merit of being entertaining.*[2]

Paine's radical argument shocked Christians who saw the credibility of Genesis as the foundation for the credibility of the Bible. Discrediting the stories of Noah's Flood and the week of Creation threatened the authority of the Bible and its promise of salvation.

Still, people on either side of the Atlantic recognized that the Bible itself had evolved, if only because the New Testament was grafted onto the Hebrew Bible (also known as the Old Testament). Long before the creation of the New Testament, after the fall of the Kingdom of David, Jewish traditions were brought together into a single history that could be handed down to preserve the cultural identity of a vanquished people exiled into captivity. It was relatively easy to accept the proposition that Jewish scribes had merged several versions of an oral tradition into a coherent whole.

A defining achievement of nineteenth-century biblical criticism was teasing Genesis apart, verse by verse, to reveal two parallel narratives. Recently developed software that analyzes style and word choices to parse authorship of multiauthored texts has found the same thing. Both the low- and high-tech methods of analysis provide support for some kind of merging of stories as a reasonable explanation for contradictions such as that between Genesis 1, in which people were created after the animals, and Genesis 2, in which Adam was created first. Such dilemmas are problematic for the simplest of reasons. At most, only one version could be correct.

Did the Flood last 150 or 40 days? In Genesis 7:24 and 8:3, the Flood is described as lasting 150 days, whereas according to Genesis 8:6–12, the floodwaters receded from the earth in just two weeks after 40 days and nights of rain (for a grand a total of 54 days of flooding). Elsewhere, the Flood was projected to last ten and a half months between Genesis 7:11, which describes the Flood as beginning on the seventeenth day of the second month of Noah's 600th year, and Genesis 8:13, which notes that the floodwaters receded enough for Noah to open up the ark on the first day of his 601st year. How can all of these things be true?

Similarly, did two or seven pairs of animals board the ark? In

Genesis 7:2–3, God commands Noah to load up seven pairs of clean animals and birds but just a single pair of the other animals. Twelve verses later, only two of each kind march aboard.

Intent on explaining such inconsistencies, biblical scholars argued that the key to disentangling the two original versions of the story lay in identifying how each version referred to God, by either the divine name Yahweh (Jehovah), or Elohim—that is, as "Lord" or "God." Perhaps the author of one version used the less formal "Elohim" because God first revealed his divine name to Moses, and so it would have made no sense to use "Yahweh" in describing the earlier history of the world. Likewise, different references to the number of animals may reflect one writer's knowledge that it was long after Noah's voyage that God revealed to Moses the distinction between clean and unclean animals.

The evidence was building for two original sources. By the end of the nineteenth century, Catholic, Protestant, Jewish, and agnostic experts alike agreed that the biblical flood story consisted of interwoven accounts fused into a single narrative during the Babylonian exile. While some biblical commentators have gone to great lengths to try to reconcile inconsistencies and apparent discrepancies, the simplest explanation for them is that earlier stories were combined. After all, how could Moses have written about his own death?

We know the New Testament was compiled from several traditions pieced together from Greek fragments, with compliers disagreeing about which books to include and which to leave out. Something similar may have happened centuries before when a newly enslaved people, desperate to preserve their oral history, wrestled over which stories to record for posterity.

The later history of the Bible shows how translation of the Hebrew word "eretz" and the Latin word "terra," both of which can mean "earth," "land," or "soil," influenced how Christians viewed topography and Noah's Flood. Saint Jerome's use of terra for both "eretz" and "adamah" (soil) in translating Genesis, and the later translation of terra as "earth," bolstered the view of Noah's Flood as a globe-wrecking deluge. But in Latin terra generally means land or soil; it does not typically imply the whole planet. The Latin word for planet Earth is "tellus." If eretz had been rendered into English in key passages as "land" rather than "earth," there might have been far less support for viewing Noah's Flood as a global event that shaped the world.

In any case, theologians have long argued that the word "earth" does not necessarily mean the whole planet. More than a century ago, conservative Church of Scotland minister Robert Jamieson pointed out that in places the Bible used "earth" to refer to limited areas, such as regions or countries. For example, God calling the dry land "earth" in Genesis 1:10 clearly implies more restricted areas than the whole planet. In other passages eretz is translated as "ground" rather than the whole planet (Judges 6:37), or as "land," when it clearly refers to a region such as the lands of Israel or Canaan (Genesis 2:11, 2:13, 13:9; Leviticus 25:9; 1 Samuel 13:3; 2 Samuel 24:8). When the same word can describe a local or regional event elsewhere in the Bible, must its use in describing Noah's ordeal necessarily refer to a global flood? Perhaps misinterpretation and quirks lie at the root of the belief in a global deluge. After all, repeated references to unicorns in the King James Bible demonstrate the potential for meanings to become scrambled as words were translated from Hebrew to Greek or Latin, and finally English.[3]

By the close of the nineteenth century, Christian theologians generally considered it reasonable to suggest that Genesis provides a synopsized or allegorical explanation of how the world came to be rather than a comprehensive history of everything that ever existed. With this simple shift in perspective, the first chapters of Genesis come into focus as the foundation for establishing a moral context for seeing the world and humankind's place in it, rather than an explanation of earth history. Reading Genesis as an epic poem intended to instruct and inspire the first monotheists rather than as a thorough blow-by-blow account of world history offers a reasonable way to solve otherwise awkward interpretive problems. But however reasonable this approach may sound, it doesn't resolve the question of where humanity's other flood stories came from—or why such stories were told all around the world.

Building on earlier work by missionaries, anthropologists had compiled hundreds of native flood stories by the early twentieth century. Missionaries, naturally, considered these tales to be degraded versions of the biblical story. Social scientists were more inclined to interpret the widespread distribution of flood stories as recording memories of prehistoric disasters, or as reflecting subconscious propensities to create flood myths. Interestingly, psychological hypotheses provide some of the most entertaining ideas. The celebrated professor of Assyriology Heinrich Zimmern claimed that the story of Noah's Flood represented a Babylonian nature myth and that "the stories of Creation, of Paradise . . . and of the Deluge all rest alike on a foundation of Babylonian material adopted by the Israelites."[4] In the authoritative 1899 *Encyclopaedia Biblica*, he maintained that the Deluge represented winter, with the Noah figure rescued in the boat standing in for the sun god. Along similar lines, the Catholic priest Ernst Böklen argued that the ark represented the moon serenely sailing across the heavenly ocean, with the moon god Noah at the helm.

After Sigmund Freud, interpretations changed. Sometimes a flood was not just a flood. One of Freud's earliest disciples, Otto Rank, described flood myths as urination fantasies. Rank went on to distinguish simple versions from those involving more elaborate birth or sexual fantasies. In his view, primitive people tended to embrace garden-variety urination myths, whereas the story of Noah's Flood represented the supreme example of a complex myth that had it all. The urinary origin of the flood was obvious enough to Rank, and to him the ark clearly represented the maternal womb, so disembarkation represented both rebirth and an invitation to procreate and repopulate the world. Other psychoanalytical approaches have also been applied to flood myths, but there is no way to either prove or disprove them—no matter how insightful or how ludicrous they may seem.

A key question is whether geology can explain flood myths and, in particular, if Noah's Flood could have been a local Mesopotamian flood that swamped the lowlands between the Tigris and Euphrates rivers. After the catastrophic floods that devastated nineteenth-century Baghdad, this possibility became far more plausible. For the most part, however, geologists avoided wading very far into biblical criticism, content to accept the premise that the story of Noah's Flood described a regional flood.

It took an Anglican bishop to push the idea that Noah's Flood was pure fiction. John William Colenso, a missionary in southern Africa who became Bishop of Natal, was greatly influenced by biblical criticism, geology, and biogeography. In *The Pentateuch and Book of Joshua Critically Examined*, published in 1864, Bishop Colenso reviewed the problems raised by believing that the flood story was true. According to the description of the Garden of Eden, the same rivers flowed in the same places both before and after the Great Flood, suggesting that Noah's Flood did little to change Earth's surface. The logistics of getting animals to and from the ark raised additional issues, as did the question of how the ark could have space for them all. But Colenso pointed out yet another conundrum. How could saving a single pair ensure the survival of species that lived in herds, like buffalo, or those that lived in hives, like bees? Without the resolution of these issues, how could people stake their spiritual salvation on belief in a global flood?

Unlike geologists such as Lyell, Colenso didn't buy the idea of a local flood. In his view, that was the easy way out. Why did birds even need to be on the ark at all when they could have simply flown off to find dry land? No, a local flood didn't make sense either. The bishop acknowledged that the Bible was clear in implying a universal flood. However, he just thought that the Bible was wrong. Noah's Flood was just a good story.

Unsurprisingly, Colenso's idea was not popular among Christian theologians. In the 1860s and 1870s, his contemporaries widely endorsed the idea of a local flood in response to the geological evidence uncovered in the first half of the century. In 1863, the authoritative *Dictionary of the Bible* dismissed the notion of a universal flood and suggested that a local flood in the lower valley of the Euphrates River provided an interpretation more compatible with geological evidence.

Secure in their faith that science and rational thought were God-given tools that could illuminate biblical interpretation, theologians from mainstream denominations acknowledged that if geology supplied evidence of only a local deluge, they would reinterpret scripture. The influential Cambridge Divinity professor Herbert Ryle expounded the belief that science was not the enemy of faith, even if the available scientific evidence required more nuanced interpretations of Genesis.

> *It must be the maxim of all reverent exposition to treat Science as the friend and not as the foe of Divine Revelation. It may be that Science seems to be but a disappointing friend when it shows the path of traditional interpretation to be no longer practicable. But the utterance of truth is the proof of purest friendship; and Science, if it closes one way, guides us to another which hitherto has been hid from view.*[5]

Ryle saw the Babylonian flood story as an ancient legend that had been incorporated into Jewish lore. The primary differences between the Babylonian and biblical stories corresponded to basic contrasts in religious thought. The moral purpose and purity of the biblical version distinguished it from polytheistic Babylonian versions. Still, Ryle saw enough differences in the narratives to think the Jews did not adopt the Babylonian flood story during their captivity. They had their own story.

In his view, the resolution to the question of how two original versions of the same story arose lay in a common ancestral tradition of a disastrous local flood that submerged the Mesopotamian world between the Tigris and Euphrates rivers.

> *There is no indication that, since man appeared upon the earth, any universal and simultaneous inundation of so extraordinary a character as to overwhelm the highest mountain peaks has ever occurred. . . . The narrative of the Flood records to us some terrible but local cataclysm which overtook the original seat of the Semitic race.*[6]

The global distribution of flood stories could be attributed to the fact that floods were common disasters all over the world.

As theologians like Ryle reconsidered traditional views, scholars began digging into the origin of flood myths and uncovered hundreds from around the globe. Most featured a hero who, like Noah, rode out the flood and repopulated the world. But there were enough differences in detail between the stories to foster debate over their origins and the question of whether they recorded a common, global disaster.

A landmark compilation of global flood traditions was included in French archaeologist François Lenormant's *The Beginnings of History*, published in 1883, which described such stories from all around the world, except Africa. He nonetheless held that most flood traditions arose from a common prehistoric event and that the Hebrew and Mesopotamian stories were identical before Abraham left for the Promised Land. According to Lenormant, India's story of Manu also came from Mesopotamia, and the Greek story of Deucalion's flood mixed the original ancient story with memories of more recent local floods. Stressing the similarities among these flood traditions, Lenormant concluded that "the Biblical Deluge, far from being a myth, was an actual and historic fact, which overwhelmed at the very least the ancestors of the three races of Aryans or Indo-Europeans, semites or Syro-Arabians, and Chamites or Kushites."[7] North American flood stories were different enough from the biblical story to preclude their having been introduced by Christian missionaries. And the Fijian flood story sounded suspiciously like a local tidal wave (what we now call a tsunami). Although the world's flood stories were rooted in fact, they didn't all arise from the same flood.

Expanding on Lenormant's study to compile a comprehensive collection of deluge traditions, Scottish anthropologist James Frazer's 1918 *Folk-lore in the Old Testament* detailed hundreds more stories of great floods. In case after case, peculiar local details appeared to be rooted in natural phenomena—a rising sea caused the flood in stories from the Pacific Islands, something Frazer thought reflected the region's history of great earthquake-generated waves. In Frazer's view, flood stories arose independently from local experiences.

Not all localities, however, gave rise to flood stories. European flood traditions were rare outside of Greece and Scandinavia. Frazer thought it remarkable that he could not find a Chinese tradition that told of a universal inundation that killed off most of the human race. Neither could he find clear cases of native flood stories in Egypt or the rest of Africa. The lack of flood stories from along the Nile—where the annual flood is quite predictable—ruled out typical river flooding as a general source of flood myths. Droughts were the real danger in ancient Egypt and along most other major African rivers where it was failure to flood that would have been catastrophic.

Frazer suggested that while Christian missionaries almost certainly introduced some flood stories, many indigenous flood stories were rooted in attempts to explain marine fossils on mountaintops or in other high places. Missionaries delighted in describing how, like Saint Augustine, native peoples around the world pointed to shells or whalebones found high on mountainsides as proof of an ancient flood.

Given the rich variety of storylines and local detail, Frazer could not see how all the world's flood traditions could be derived from the biblical story. In contrast, Frazer thought it was easy to see how local stories of catastrophic floods would evolve into stories of a universal deluge.

> *On the whole, then, there seems to be good reason for thinking that some and probably many diluvial traditions are merely exaggerated reports of floods which actually occurred, whether as the result of heavy rain, earthquake-waves, or other causes. All such traditions, therefore, are partly legendary and partly mythical: so far as they preserve reminiscences of floods which really happened, they are legendary; so far as they describe universal deluges which never happened, they are mythical.*[8]

After Frazer's exhaustive study, only those uncritically seeking to legitimize a global flood gave any credence to the argument that the global distribution of flood stories meant they shared a common origin.

Those still trying to argue that the global distribution of flood stories is a legacy of a global flood have to consider how the rich collection of Chinese flood stories has very different storylines from Mesopotamian flood stories. Historian Mark Lewis's *The Flood Myths of Early China* relates how the storylines and themes of Chinese flood stories are strikingly different from those at the roots of Western culture in presenting prevention of the flood as a human triumph. They do not tell of divine vengeance and human frailty but demonstrate how human labor can overcome nature.

Interpretations of Chinese flood stories point to their use as effective sanctions for traditions, laws or institutions in describing the construction of order from the chaos of a universal flood. In some versions, those who caused the flood are described as hooligans challenging the proper order of things. Such stories provided a charter for early imperial institutions, relating how the flood dissolved the distinctions between different classes of men—with disastrous consequences. Some versions of Chinese flood stories focus on failed attempts to impound the floodwaters, emphasizing the importance of flood control, which, in turn, helped justify the authority of rulers who maintained the all-important levees.

The savior-hero Yu is the central figure of Chinese flood myths dating as far back as 1000 BC. He drained lowland floodwaters so that the fields could be planted and dredged rivers so that they could be channeled to the sea. This divided the world (China) into natural provinces. In one version, Yu is described as the minister of works who "stabilized the water and land,"[9] setting the stage for the arrival of agriculture and the development of Chinese civilization. His work is credited with allowing people to "descend from the hills and dwell in the fields,"[10] something that parallels how Chinese society moved from the eroded uplands on the edge of the Tibetan Plateau down to farm the fertile floodplains.

One version of the story holds that in the time of the flood the world was covered by wild grasses and forests inhabited by birds and wild animals. Taming the floodwaters allowed for the domestication of crops and the expansion of human settlement, bringing order to the land. These accounts of subduing nature's chaos sound like draining swampy lowlands to transform wild land into farmland. The contrast with Mesopotamian stories of a killer flood sent by a vengeful god is striking.

Deciphering the origin of many flood stories is complicated because the biblical story may have hybridized with native legends. Noah's story is one of the most colorful in the Bible and would have impressed people with a flood tradition of their own. But because missionaries were often the first to record local stories, it can be hard to tell whether a flood story predates Christian contact or just regurgitates Noah's story with local color added.

Anthropologist Alice Lee Marriott inadvertently discovered how rapidly stories can jump from one culture to another while collecting Native American folklore in South Dakota in the summer of 1936. One day an elderly informant challenged her to tell him one of her people's tales. She told him a version of Beowulf as the story of a brave warrior and the water monster. Afterwards, she was impressed with how he rounded out details to improve the story in retelling it to his people. A few years later Marriott was amused to find her story as the subject of a research paper in an ethnological journal documenting a Beowulf-like myth among Native Americans.

A century before, in 1842, a missionary named Moffat told the tale of how he could not find a flood legend among South Africans until one of the Khoikhoi (whom colonists called Hottentots) told him the story of a great flood. The man assured Moffat that this was a tale of his forefathers, and that Moffat was the first missionary he had ever met. Later, in comparing notes with another missionary, Moffat learned that his colleague had indeed told his native informant the story of Noah's Flood. This shows how difficult it is to determine the origin of many flood myths due to the potential for unrecorded cultural transmissions.

Unsurprisingly, people living in flood-prone estuaries are likely to have stories of a great flood. The estuary of the Tigris and Euphrates rivers receives its water from the mountains of Turkey and Iraq, and a warm spring rainstorm falling onto a heavy snow pack can submerge the whole floodplain under many feet of water. When the levees burst there is nowhere to go as everything slips under water. Every now and then people living in this region were forced to flee to higher ground or pack their possessions and animals onto a boat or raft as their world sank beneath floodwaters. The lack of well-documented flood myths from Egypt and the Nile River may be due to the fact that the Nile gets its water from sources far to the south in equatorial Africa. Fed by a chain of great lakes in the East African Rift, the river's annual discharge does not vary anywhere near as much as in Mesopotamia. The predictably moderate annual flood was no threat, it was the source of life.

How long could stories of a great flood survive oral transmission from one generation to the next? Examples of stories that have been passed down through oral transmission for thousands of years have been reported from several continents. My favorite is a Klamath Indian story, recorded in 1865. It provides a compelling eyewitness account of the eruption of Mount Mazama, which formed Oregon's Crater Lake more than 7,600 years ago. For tens of thousands of years, our preliterate ancestors conveyed knowledge from one generation to the next through oral traditions. For a story to survive retelling over many generations it has to be viewed as important, it must continue to have relevance or relate to something still visible to listeners, and it must be highly memorable. Stories of a great flood satisfy all three criteria, particularly in flood-prone regions.

Upon reflection, my theory that flood stories from around the world are grounded in reality is plausible. For tens of thousands of years, oral traditions were the only means of transmitting information from one generation to the next. And while not all stories bear retelling, tales of disastrous, displacing floods were sure to be retold for generations. Just think of your own family's lore. It's not the day-to-day events that get passed on, it's the big, memorable things.

After the devastating blows to flood geology in the first half of the nineteenth century, geologists increasingly avoided debates over how to account for the biblical flood. The educated consensus was that just because it was written for an audience with a Mesopotamian knowledge of earth science didn't mean that the Book of Genesis wasn't written to convey the majesty, scope, and power of creation.

By the end of the nineteenth century, mainstream geologists had lost interest in the Deluge. It was a settled matter. Noah's Flood was widely seen as a local historical event in the Middle East, even if its precise nature remained debatable.

Thomas Huxley, the last survivor of the generation of prominent scientists who lived through the battles over Lyell's and Darwin's work, even wrote an essay arguing that a global deluge inundating the world was a fable that conflicted with geological evidence. He recalled the century's changes in the relation between geology and Christianity:

> *At the present time, it is difficult to persuade serious scientific inquirers to occupy themselves, in any way, with the Noachian Deluge. They look at you with a smile and a shrug, and say they have more important matters to attend to. . . . But it was not so in my youth. At that time, geologists and biologists could hardly follow to the end of any path of inquiry without finding the way blocked by Noah and his ark, or by the first chapter of Genesis; and it was a serious matter, in this country at any rate, for a man to be suspected of doubting the literal truth of the Diluvial . . . history.*[11]

Huxley virtually credits Lyell with single-handedly creating the science of geology, ignoring the contributions of Buckland, Sedgwick, and others who also struggled with and turned against the idea of a global flood. Perhaps Huxley relegated them to the background because of their membership in the clergy, the villains of his story. Huxley's portrayal of a century-long battle between Lyell's rationalism and blind faith in a catastrophic global flood fostered the perception of an ages-old war between Christianity and science.

At the dawn of the twentieth century, geologists were almost entirely uniformitarians. Lyell's dictate that the present was the key to the past had become geological dogma. A growing body of geological evidence and alternative explanations for Siberian mammoth carcasses effectively dismantled the remaining fragments of a case for a global deluge as the primary driving force in earth history. But over the course of the twentieth century, the rise of flood geology proponents among evangelical Christians fostered the view that geology and faith—science and religion—could not peacefully coexist. Instead of trying to refine their understanding of the biblical flood story in light of new knowledge, radically conservative Christians broke with those who acknowledged scientific findings and began to ignore, selectively cherry-pick, and actively undermine science to support their favorite literal interpretation of the Bible. Today, we know them as creationists.

10

Dinosaurs in Paradise

When I heard that the new Creation Museum in Petersburg, Kentucky, featured exhibits showing people picnicking with dinosaurs in the Garden of Eden, I had to see it. Nothing could have prepared me for a dinosaur-petting-zoo version of natural history. Upon entering, I was greeted by a diorama showcasing a velociraptor straight out of *Jurassic Park* calmly standing beside Eve while she feeds a squirrel.

Visitors pass a ticket checker dressed up as a Park Ranger stationed at a Grand Canyon National Park sign, then navigate a fake bedrock canyon designed to enthrall kids and arrive at a large two-panel board that addresses the issue of the age of the universe. The left-hand side says that reason holds the universe to be billions of years old. The right-hand side indicates that God says that it all began six thousand years ago. So which should we believe—reason or God, the creator of reason?

I was prepared for unusual perspectives, but one of the next panels caught me off guard by endorsing evolution. Its diagrams illustrated several versions of the tree of life to contrast the scientific view with the creationist view of what really happened. Alongside the conventional portrayal of life evolving from single-celled organisms to modern flora and fauna, the display illustrated how a limited number of species in God's original "creation orchard" started branching into new species before Noah's Flood. Afterward, some, like dinosaurs, went extinct, while their luckier peers rapidly flowered into modern species. The diagram for humans stood out as a simple straight line, showing no change from Creation to the present.

More surprises awaited me down the next hall, where floor-to-ceiling panels asserted that scientists throughout history conspired to question, destroy, discredit, criticize, poison, and replace God's Word. In this view, the dangerous brotherhood of science is humanity's common enemy. Reason threatens us all.

After absorbing the anti-reason display, visitors advance to the modern world through a graffiti-filled alleyway, where mock windows voyeuristically display videos of a teenage boy watching pornography and a girl seeking an abortion. Across the alley a wrecking ball demolishing a church is branded with giant letters spelling out "millions of years." The message is clear. Belief in geologic time drives the decay of modern society.

Moving through the next display, a Garden of Eden diorama where people and dinosaurs frolic together and signage says carnivores didn't eat meat, I came to the creationist perspective on geology. The exhibit told how Noah's crowded ark surfed a great wave that swept back and forth across the world. After the world-remodeling Flood, nothing much happened, except for a few volcanic eruptions and earthquakes scattered here and there throughout history. That rivers and glaciers could sculpt topography is summarily dismissed as the deranged product of human reason.

In this depiction, geologic time never happened. Gone are centuries of painstaking work to piece together the story of our planet.

Gone are the overlapping tree-ring records that meticulously matched up patterns of annual growth to reach back more than ten thousand years. Gone are the hundreds of thousands of individual layers recording annual snowfall recovered from cores drilled through the polar ice caps. Gone are the revelations of plate tectonics that elegantly tied earth history together in a unifying framework, explaining the form of continents and their wanderings over millions of years. Gone, in fact, is nearly all of earth history.

In addition to the inherently untestable idea that a divine being created the universe with a particular plan in mind, creationists advocate testable interpretations of earth history. Because their ideas have failed when put to the test, they declare reason to be their enemy.

Even minimal geologic training equips one to see how the material displayed in some of these exhibits contradicts the interpretive signage. For example, dinosaur tracks preserved in layers of sedimentary rock present a serious problem for creationists. How could land animals have been walking around on the seafloor during an event that ripped up Earth's surface before depositing their bones in the very stuff they were walking around on? Likewise, it is readily verifiable that it takes more vigorous flow to erode hard bedrock than to deposit loose sediment. How, then, could the peak of the Flood have laid down all the sedimentary rocks before the waning stages ripped open the Grand Canyon and carved out the world's topography?

And why does this museum have so many displays showing giant reptiles hanging out with Adam and Eve when the Bible doesn't even mention dinosaurs? Because if Noah's Flood is pretty much all there was to earth history since the Creation, then dinosaurs must have lived alongside people in the days before the Flood. How did such beliefs gain traction?

We can trace the roots of modern creationism back to the nineteenth century, when geology emerged as a profession distinct from theology and natural philosophy. As geologists abandoned Noah's Flood as a central subject and moved on to other pursuits, Christianity splintered into those willing to accept geological findings and those who insisted on the reality of a global flood. The later conflict over evolution served to strengthen such differences. As mainstream Protestants and Catholics adapted biblical interpretation to accommodate geology, a new breed of American fundamentalists defended the reality of a world-destroying flood as central to their faith.

The Bible was one of the only traditional sources of authority that emerged from the American Revolution unscathed (despite the best efforts of Thomas Paine). The war fostered independence in multiple forms and encouraged the revolutionary conviction that everyone (except women and slaves) possessed both common and moral sense. American Protestants began rejecting traditional forms of authority, confident their own vision would lead them closer to God. This commonsense populism paved the way for the fundamentalism that, in turn, spawned modern creationism.

In the early nineteenth century, camp meetings and revivals brought organized religion along as westward migration took people far from the established churches of the eastern seaboard. One of the first, Kentucky's Cane Ridge Revival of 1801, was attended by thousands eager to hear populist preachers, gamble, and carouse—not necessarily in that order. The popularity of the weeklong meeting taught frontier preachers a winning strategy for spreading the Gospel across America.

In contrast to Presbyterian denominations that disciplined ministers who participated in boisterous revivals, Methodists and Baptists used the rowdy meetings to swell their ranks. Employing charismatic preachers with little or no education who could relate to the masses heading west, these sects grew into the largest Protestant congregations by the close of the frontier.

Populist preachers who considered the common sense of ordinary men more reliable than opinions espoused by seminary-trained theologians and book-learned professors encouraged people to cast off the chains of religious authority and interpret the Bible for themselves. The most successful preachers—those whose flocks grew the fastest—adopted popular language and manners. When coupled with belief in the Bible as the sole source of religious authority, populism encouraged settling theological disputes in the court of public opinion where everyone was entitled to interpret the Bible for him- or herself. This produced an interpretive free-for-all in which discredited ideas could compete with reasonable ones.

Sectarianism flourished in America's religious marketplace. Splinter groups left mainstream denominations in disputes over doctrine, practice, and/or belief. Although the founders of these new denominations obviously disagreed on matters important to them, most shared the belief that the Bible was the only real authority for Christians and that its meaning was laid out plainly. Scripture meant exactly what it said, even if they didn't agree on what it meant.

The advent of the American Civil War presented a theological crisis for American Christians. Both North and South used the Bible to either condemn or defend slavery. How could a plain-sense interpretation of scripture be infallible if one side had to be wrong? Such dilemmas only hardened divergent interpretations of the Bible.

Conservative Protestants began to forge a reactionary biblical literalism, based on biblical inerrancy. They believed that admitting even the slightest error in or sign of human influence on the sacred text would undermine the whole notion of Christian salvation. One need not look for deeper meanings because common sense tells us what the Bible means. Efforts to uphold literal, plain-sense scriptural interpretations began to distance evangelicals from mainstream thought.

Fundamentalism arose among conservative Protestants who viewed liberal accommodation of modern ideas and values as a betrayal of the core doctrines they viewed as fundamental to their faith. Foremost among these was biblical inerrancy. In 1895, the founding fathers of fundamentalism declared this doctrine one of the "five points of fundamentalism" at the Niagara Bible Conference where they staked out their unnegotiable beliefs. Two decades later, the conservative Protestant academics who authored *The Fundamentals*, a series of essays published between 1910 and 1915 that gave birth to fundamentalism, attacked critical historical and literary analysis that questioned biblical authority.

At first fundamentalists did not insist on strict biblical literalism. The Bible could not be wrong, but interpretations could adapt as needed to preserve biblical infallibility. The Bible could be read in different ways. The original fundamentalists juggled what to read figuratively and what to read literally in order to preserve biblical infallibility. Their approach was surprisingly flexible in comparison to their counterparts today. Most accepted an old Earth through either the day-age theory or the gap theory and were open to the idea that Noah's Flood may have been a local affair that wiped out humanity's roots.

By the 1920s, a loose coalition of militant Protestants began to characterize liberals as false Christians who had lost faith in traditional beliefs and doctrines. Claiming to defend the true faith, newly militant fundamentalists combined biblical inerrancy with biblical literalism. Their zeal to combat biblical criticism lay in the conviction that admitting the Bible had a history colored by human fallibility opened the door to doubting redemption through Christ. A literal reading founded on biblical inerrancy formed the levee fundamentalists built to save the Bible from the flood of modernism.

Fundamentalists became increasingly isolated as their efforts to stem the rising tide of liberal thought failed to sway mainstream denominations in the 1930s. They then focused on building their own network of churches and schools dedicated to teaching biblical infallibility. As fundamentalists began slipping into a self-contained world, the recycled arguments of flood geology seemed to provide fresh ammunition for the fight to ban teaching evolution in public schools—and its heretical foundation in an ancient Earth.

By the mid-twentieth century, conservatives militantly pushing literal biblical interpretation stopped interacting with geologists just as breakthroughs like the ability to use radioactive decay to directly date the age of rocks and fossils began to revolutionize the earth sciences. Paleontologists, in particular, threw cold water on the creationist idea that mammoths were flash-frozen or buried in a sudden environmental calamity.

In 1929, Carnegie Museum curator of paleontology Innokenty Tolmachoff meticulously described the circumstances and condition of every known mammoth carcass discovery dating back to the seventeenth century. Three dozen sites pretty much accounted for them all. Noting evidence that mammoths ate great volumes of tundra grass in the summer, Tolmachoff lambasted claims that mammoths roamed a more temperate Siberia. Mammoths were creatures of the ice age, not victims of it. They only went extinct at the end of the last glaciation.

Tolmachoff also reported that stories of mammoth carcasses preserved well enough to eat were greatly exaggerated. Dogs greedily devoured thawed mammoth, but people found it inedible. As far as he could tell, there was no basis for tales of great feasts prepared from their frozen carcasses. Firsthand accounts consistently reported putrid flesh in advanced states of decay. And the circumstances surrounding their discovery suggested that mammoths became stuck in soft mud, were caught in collapsed thawing ground, or drowned along big rivers. They died mundane, solitary deaths.

Such evidence did not dissuade the followers of George McCready Price, a prolific, self-taught writer of geology books, despite having no geological education or training. The writings of Ellen Gould

White, founding prophetess of Seventh-day Adventism, convinced Price of the validity of flood geology. He rejected the popular day-age and gap theories based on White's accounts of visions she'd had in which she saw God create the world in six twenty-four-hour days and rest on the seventh. Her trancelike visions revealed that fossils were buried when Noah's Flood reworked Earth's surface. Explaining how God removed all the rotting carcasses after the Flood, she told of how a great wind carried "away the tops of mountains like mighty avalanches, forming huge hills and high mountains where there were none to be seen before, and burying the dead bodies with trees, stones, and earth."[1] All that buried vegetation turned into coal, which God occasionally ignited when He wanted to fire up volcanoes. White's fantasylike explanations sound like the wild ideas of seventeenth-century natural philosophers.

Born in rural New Brunswick in 1870, Price was a child when his father died and his mother joined the apocalyptically inclined Adventists. Fresh out of high school, he married an older Adventist woman, and together they made their living selling White's Adventist books door-to-door across Canada. A few years later, in 1891, Price enrolled in Battle Creek College, an Adventist school in Michigan, but fell back to selling books two years later when his money ran out.

Around the turn of the century, when serving as a high school principal in eastern Canada, Price nearly succumbed to the local physician's views on evolution after borrowing volumes from his friend's library. Price concluded that a solid geological foundation would make evolution appear to be reasonable. He came close to accepting that there really must be something to the idea of vast geological ages and worlds lost to the depths of time. But how could he reconcile geologic time with White's teachings? Guided by prayer, he decided that geologists were fooling themselves. Fossils were really all the same age. Shocked by how he almost yielded to temptation, Price vowed to promote White's vision of how Noah's Flood accounted for the fossil record. He had at last found his calling.

Several years later Price had ample time to ponder how to refute geological theories while working as a handyman at an Adventist sanitarium in southern California. In 1906, his self-published and aptly named *Illogical Geology* attacked the geological foundations of evolution and claimed there was no proof that any fossil was older than any other. The succession of organisms that geologists found in the rocks was really a mixed-up sampling of communities that lived in different parts of the world before the Flood. What really happened was that a sudden shifting of Earth's axis had released great subterranean reservoirs and drowned the world. Then a miraculous cosmic storm buried all the drowned bodies and kept the atmosphere from going putrid. Afterwards, the receding waters carved natural wonders like Niagara Falls and the Grand Canyon. His geological story reheated Burnet's and Woodward's stale theories.

Price sent copies of his book to eminent geologists seeking their reaction. Among the few who bothered to respond was David Starr Jordan, president of Stanford University and an expert on fossil fishes. In a letter to Price, Jordan warned him not to expect geologists to take him seriously because his argument was based on "mistakes, omissions and exceptions" that rendered his case "as convincing [as] if one should take the facts of European history and attempt to show that all the various events were simultaneous."[2] Equally impressed by Price's obvious intelligence and ignorance of geology, Jordan tried for over two decades to convince him to get some experience in field or laboratory work. Decades later, students on a fossil-hunting trip were astonished to discover that the world's leading creationist could hardly tell one fossil from another.

The roots of modern creationism run directly back to Price. Honing arguments faithful to White's teaching, Price convinced himself that it was the theories of geologists and not the rocks themselves that opposed a literal reading of Genesis. He called his view of geology the new catastrophism to distinguish it from earlier views of earth history involving multiple catastrophes.

Initially, Price made little headway among fundamentalists and he was careful not to point out the incompatibility of his views with the widely accepted day-age and gap theories. Most fundamentalists committed to scriptural inerrancy followed the conservative *Schofield Reference Bible*, which endorsed the gap theory in explaining that the original Creation in the first verse of Genesis "refers to the dateless past, and gives scope for all the geological ages."[3] Price was a lonely voice insisting on the literal truth of a global flood that rearranged Earth's surface and deposited the whole fossil record along with all the world's sedimentary rocks.

Geologists ridiculed his ideas mercilessly. Professors routinely assigned graduate students the exercise of refuting them. Writing in *Science* in 1922, Arthur Miller, the head of the geology department at the University of Kentucky, described Price as an "alleged geologist . . . who, while a member of no scientific body and absolutely unknown in scientific circles . . . is hailed by the 'Fundamentalists' as their great champion—one who . . . has brought into prominence the 'heretofore mute evidence of a mighty upheaval and a flood.' "[4] Miller was amazed that Price had the audacity to accuse geologists of being biased when Price's new catastrophism "turns out to be nothing more than the Old Catastrophism embodied in the Noachian Deluge."[5]

When Price read Miller's disparaging remarks, he fired off an angry letter threatening to sue if not given the chance for a rebuttal. The editor offered to correct any errors of fact but declined to publish Price's geological views. In response, Price unleashed a furious retort in the *Sunday School Times*.

Convinced that a great flood remodeled the entire world, Price called on vast mammoth graveyards as evidence of a sudden calamity, unaware that none had actually been found. He repeated the apocryphal stories of frozen mammoth proving fresh enough for a feast, apparently unaware that firsthand reports contradicted this popular misconception. He also argued that coal deposits and fossil coral found at high latitudes indicated a warmer pre-flood world. He considered this last point particularly persuasive because geologists could not yet explain the fossilized remains of tropical organisms found near the poles.

Price published *The New Geology* in 1923, covering standard introductory subjects. Written to look like a textbook, although aimed at the general reader, Price's book attacked conventional notions of geology. The uninformed reader would see nothing in it to indicate that this did not lay out the essentials of modern geology. Until, that is, one discovered Price's assertion that geological understanding of a progressive succession of organisms through geologic time was not only flawed, but had been "disproved by a large number of recently discovered facts" that he neglected to mention.[6] Instead, he simply asserted that all the animals in the entire fossil record—trilobites, ammonites, dinosaurs, and mammoths—lived together in harmony with people before the Flood.

Whether ignorant or simply dismissive of centuries of discovery and debate, Price attributed the entire geologic record to Noah's Flood depositing enormous piles of sediment chock full of fossils. Settling disrupted the pile where the basement strata were unable to support the extra load. Arguing that the folding and tilting of rocks occurred while they were still soft, Price accused mainstream geologists of raw prejudice as he himself never bothered to learn any geology and ignored evidence accumulated by generations of geologists.

Isolated from contact with geological thinking, fundamentalists looking for arguments to use in their attacks on evolution in the 1920s turned to Price's flood geology, trusting that it was based on sound science. With no trained geologists among hard-core evangelicals,

Price was virtually unchallenged as the sole geological voice in fundamentalist ranks. Offering a message right on target for the war on evolution, Price became a fundamentalist darling. By the mid-1920s he was a regular contributor to conservative religious periodicals. In short order, although he had no scientific background or training, he became the fundamentalists' principal scientific authority.

Fundamentalist beliefs on evolution came to a head in the spring of 1925, when high school teacher John Thomas Scopes confessed to violating a state law against teaching human evolution in Dayton, Tennessee. At his famous trial, defense attorney Clarence Darrow called prosecutor William Jennings Bryan to the witness stand as his final expert on the relation between science and the Bible. Bryan was a well-known politician who jumped at an opportunity to campaign against the moral decay that set in when evolution encouraged people to question biblical authority.

Darrow grilled Bryan about a host of biblical absurdities. Where did Cain, the murderous son of Adam and Eve, find his wife if his parents were the only other people on Earth? Was Jonah really eaten by a whale and then spit up alive after spending days submerged in the belly of the beast? How could Bishop Ussher's 4004 BC date for the creation be accurate when Chinese and Egyptian history extend back farther in time? Could Bryan point to any credible scientist who believed that the story of a global flood could be taken literally? In response, Bryan named Price.

Hearing this, Darrow scoffed, "You mentioned Price because he is the only human being in the world so far as you know that signs his name as a geologist that believes like you do . . . every scientist in this country knows [he] is . . . a pretender and not a geologist at all.[7]" Darrow went on to get Bryan to admit that the days of Genesis 1 were not literal twenty-four-hour days. Each day might have lasted for millions of years. The planet itself might be quite ancient even if people were created just six thousand years ago. Although Bryan reportedly believed in a local rather than a global flood and equated young-Earth creationists with flat Earthers, it did not stop him from using Price's flood geology to attack evolution.

At the end of the day, despite Bryan's joking rejoinders, Darrow had made his point that literalists interpreted the Bible as much as anyone, cherry-picking their way through Scripture. The other defense attorney, Dudley Field Malone, noted that Bryan's reading was not the only way for Christians to interpret the Bible: it was possible to accept modern science as not being at odds with religious truths.

The press was not at all kind to Bryan. Neither was fate. He died right after the trial.

Creationists changed tactics and turned on librarians and teachers, harassing them to keep textbooks that fundamentalists considered objectionable out of classrooms. Creationists who had made front-page headlines in the 1920s were all but forgotten a decade later. Shut out of the popular press, they turned to building their own institutional base, starting their own organizations, journals, and schools. Fundamentalists of this era varied greatly in terms of what to believe about geological ages and the biblical flood. Some, like Price, held to the strict literal interpretation of six days of creation followed by a global flood. Others promoted the gap theory or the idea that each day in the week of creation represented a whole geological age. Leading fundamentalists began to wonder how evangelical Christians could convert the world to their views if they didn't even agree among themselves.

Of course, when Price first claimed that all the organisms preserved as fossils died in a sudden catastrophe, there was no way to date their deaths and directly test his claim. Steno's approach could reveal the relative age of the geological formations containing fossils by determining their order of deposition, but there wasn't yet any way to directly measure the age of fossil-bearing rocks or fossils themselves.

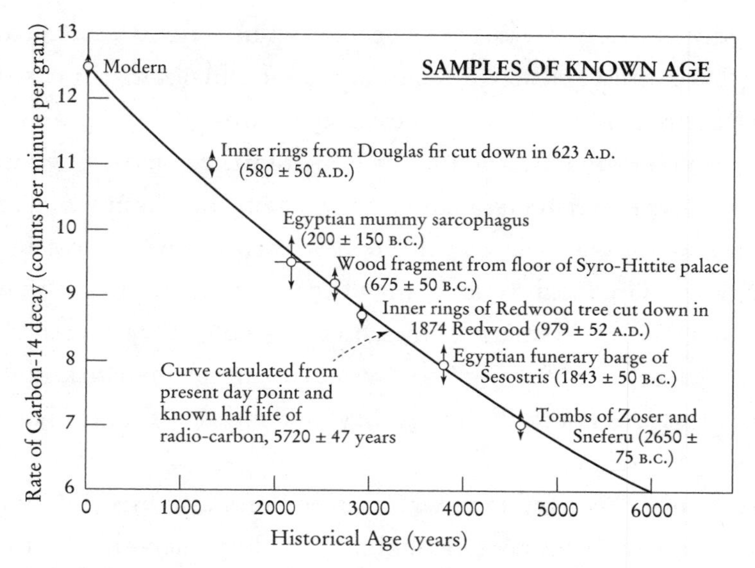

Graph showing first test of radiocarbon dating in a plot of known sample age versus the rate of carbon-14 (^{14}C) decay and the close fit between measured values (data points) and values predicted (curve) by radiocarbon decay (based on a figure in Arnold, J. R., and Libby, W. F., 1949, Age determinations by radiocarbon content: Checks with samples of known age, Science, *v. 110, p. 678-680).*

The development of radiocarbon dating was revolutionary, as it enabled scientists to reliably date deposits from the last few tens of thousands of years. The method was developed by Willard Libby at the University of Chicago's Institute for Nuclear Studies and is based on measuring the rate of decay of the naturally occurring unstable radioactive isotope ^{14}C (carbon-14). Collisions between cosmic ray protons and particles in the atmosphere produce secondary neutrons that are captured by nitrogen nuclei in the N_2 gas that forms most of the atmosphere. This fusion creates ^{14}C, which decays to the normal stable nitrogen isotope (^{14}N) with a characteristic half-life of about 5,720 years, the time it takes for half of the amount remaining to decay. When plants convert atmospheric carbon dioxide (CO_2) into organic matter during photosynthesis, a small amount of ^{14}C

is incorporated in proportion to the amount in the atmosphere. The atmospheric ratio of ^{14}C to ^{12}C is maintained in living things that continually incorporate new carbon into their bodies. But after they die, the ^{14}C no longer gets refreshed and starts to decay exponentially—at a rate proportional to the amount left. Libby reasoned that if one knew the half-life of ^{14}C, one could tell how long decay had been going on by measuring the present rate of decay.

He tested the technique by dating wood from samples with a range of independently known ages. The youngest came from a piece of Douglas fir cut down in 623 AD. Others included the sarcophagus of an Egyptian mummy dating from the third century BC, the inner rings of an almost three-thousand-year-old redwood tree, deck boards from the funerary barge of an Egyptian pharaoh who died around 1843 BC, and wood from a pair of five-thousand-year-old tombs. The ages predicted by radiocarbon dating closely agreed with the known ages of the samples. Radiocarbon dating worked.

Its application to woolly mammoth carcasses presented a serious problem for champions of flood geology. Carbon dating showed that mammoth carcasses range from more than forty thousand to less than ten thousand years old, disproving the single catastrophe theory. Mammoths did not all die at once.

How did evangelicals respond to these findings? Many accepted radiometric dating, the idea of an old Earth, and the possibility of a regional flood. But those fundamentalists committed to flood geology and a young Earth responded not with facts or a reinterpretation of scripture; they simply refused to believe it.

This didn't solve their mammoth problem. Studies of individual mammoth carcasses revealed that mammoths did not all drown, as they surely would have in a global flood. Some died in the old-elephant death position, down on the stomach with legs stretched out in front. Others sank through the permafrost, fell into collapse pits, or got stuck in swampy ground, unable to extract their bulk from the mire. Mosses, grasses, and herbs found in mammoth stomachs were characteristic of the vegetation growing within a few hundred kilometers of their carcasses. There was no need to invoke a global flood to deliver them from the tropics. Mammoths lived and died close to where their remains were found.

None of the arguments for asserting that mammoths died in a great catastrophe survived twentieth-century scrutiny. Creationists didn't seem to notice.

Concerned over growing antagonism toward science in their community, evangelical Christians formed the American Scientific Affiliation (ASA) in 1941 to promote study of the relationship between science and the Bible. One of its key members was J. Laurence Kulp, a PhD chemist from Princeton University who had mastered radiocarbon dating in Libby's lab at the University of Chicago. He went on to become a leading authority on the method and established his own carbon dating lab at Columbia University. In an article published in 1950 in the *Journal of the American Scientific Affiliation*, Kulp attacked flood geology as an embarrassment to both science and Christianity.

Kulp's influence helped split the ASA into two camps: old-Earth believers and young-Earth creationists. The former believe that God created the world, but at a geological pace. Bitter disagreements grew into a rift that still characterizes evangelical Christianity today as young-Earth creationists began attacking the idea of an old Earth that allowed time for evolution.

Kulp noted basic creationist errors that he thought reflected a lack of education and training among prominent advocates of flood geology, especially in the important subdisciplines of field mapping, paleontology, and structural geology. Creationists held that geology and evolution were synonymous even though the geological basis for determining the relative age of rocks did not actually rely on fossils. Creationists also claimed that the conditions under which rocks formed and deformed were not well understood. Kulp attributed the confidence of flood geologists to their sincere belief in these fallacious convictions. Finally, he charitably maintained that flood geologists were simply out of date. They relied on Price's work, which predated the development of radiometric dating, the perforation of Earth's sedimentary cover by oil wells, and studies that conclusively documented the conditions under which sedimentary rocks form and deform. In other words, so-called flood geologists simply didn't know what they were talking about.

In debunking flood geology, Kulp focused on the formation of sedimentary rocks, pointing out how it was impossible for them to have all formed during a single flood. In the 1930s, cores from Venezuelan oil wells documented a complete section showing the compaction and transformation of river mud into hard shale. Penetrating through two vertical miles of muddy sediments, the drill cores revealed that loose mud had to be buried under at least a mile of sediment before it solidified into rock. A mile of water would not do the trick because the additional weight of overlying sediment was needed to squeeze water from the mud. Similar studies documented comparable results for limestone and sandstone. If sedimentary rocks now exposed at the surface all formed during the Flood, then where did the mile of sediment that must have covered them go if there was only a few thousand years to erode it all off?

Even more damning was Kulp's discussion of the problem of how to warp layers of sedimentary rock into broad regional folds like those that characterized Appalachian geology. Creationists attributed such deformation to the slumping of Flood-deposited mud and sand, before these layers hardened into rock. Kulp described how this was physically impossible. Shell Oil Company geologists had shown that in order to reproduce geologic conditions in a laboratory setting, one had to scale all the dimensions in the model—including the material properties. Using modeling clay to experimentally investigate the deformation of rocks at temperatures and pressures equivalent to about five to ten miles down within Earth's crust, one could easily reproduce the folding seen in sedimentary rocks. While turning loose sediment into solid rock required burial to considerable depth, folding rocks required even deeper burial and higher temperatures. Flood geology simply could not explain the world's great expanses of folded sedimentary rock.

Kulp also described how radiometric ages of rocks determined by measuring uranium-lead ratios agree with the stratigraphic order worked out by field geologists on the basis of Steno's principles for interpreting structure and stratigraphy. Radiometric dating confirmed the basic order to the stratigraphic record independently from the fossils sedimentary rocks contained. Price's argument that geologists used the idea of fossil succession (and thus evolution) to impose an artificial order on the geologic record showed how little Price understood geology.

Kulp asked how if sedimentary rocks really were deposited by great waves moving at speeds up to a thousand miles an hour it would be possible to preserve the kind of ecological zonation creationists called upon to explain fossil assemblages—the idea that the different fossils that characterized different rock formations simply reflected the animal communities in different ecological zones on the pre-Flood Earth. Such a violent current would mix and remix anything ripped up from Earth's surface. The ecological zonation that creationists invoked to explain the fossil record could not survive the flood they called upon to generate it.

One of the simplest arguments against a young age for the world's sedimentary rocks was the amount of water that would need to be evaporated in order to account for the great thickness of evaporites, like the gypsum (calcium sulfate) deposits in Michigan and west Texas. Since less than a foot of gypsum would precipitate out of a thousand feet of seawater, Kulp calculated that it would take evaporation of an ocean 450 miles deep to build up the thick gypsum deposits of west Texas. Based on the most extreme recorded evaporation rates from the Dead Sea, he calculated this would take hundreds of thousands of years. The world's thick evaporite beds could not have formed in the single year of Noah's Flood.

Evidence based on completely different approaches—radioactive decay, the amount of salt in the sea, and even the relationship between the speed of light and the distance to the stars—all indicated that Earth was millions if not billions of years old.

Kulp concluded his critique of Price's ideas by warning that pushing demonstrably false ideas would hinder the spread of the Gospel among educated people. An evangelical himself, Kulp studied chemistry until he felt the Lord call him to study geology. He was concerned that for half a century too few evangelical Christians had entered the field of geology; consequently, Price and his disciples exercised too much influence in evangelical circles, given their lack of geological knowledge.

Few mainstream Christian scholars bought into Price's flood geology. In 1954, influential Baptist theologian Bernard Ramm critiqued creationism from an evangelical perspective in *The Christian View of Science and Scripture*. Ramm argued against a recent global flood. He considered it ludicrous to think that people from all the world's ethnicities could have descended from Noah in just a few thousand years.

Ramm contrasted two traditions through which Christians approached science. Those adopting the "ignoble tradition" had taken a hostile attitude toward science and "used arguments and procedures not in the better traditions of established scholarship," whereas those following the "noble tradition" had "taken great care to learn the facts of science and Scripture."[8] To set science against religion was to set creation against creator. "If the Author of Nature and of Scripture are the same God, then the two books of God must eventually recite the same story."[9] Ramm advised evangelical Christians not to confuse interpretation with revelation. Just because the Bible was the infallible Word of God did not mean that it was always obvious as to what it meant regarding scientific matters. Confidence that one understood the clear meaning of scripture did not necessarily mean one did.

In defending radioactive dating of rocks, Ramm related how experiments under a wide range of pressures and temperatures showed no effect on the rate of radioactive decay. Radioactive isotopes changed at a constant rate. Geologists could tell how long a sample of uranium (or carbon) had been decaying in a similar way to how we could "measur[e] how much gas we have left in the tank [to get] an idea how many miles we have driven."[10]

For Ramm, the idea that Earth existed for millions of years before God reconditioned it for human use adequately reconciled Genesis and geology. In the epilogue to his book, Ramm pointed out that not only did evangelicals of his day not believe that Earth is either flat or at the center of the universe but that many considered the findings of modern geology to be perfectly consistent with their faith.

Ramm's book caused quite a stir among fundamentalists. A leader of the self-described new evangelicals, he sought to engage modern culture, avoided belligerency, and embraced scholarship. Shortly after it was published Billy Graham praised Ramm's book and called for a new view of biblical inspiration that respected and accommodated modern science. It seemed as though the idea of a global flood was vanquished. No serious scientist or mainstream theologian still gave it any thought. The key to accepting the fact that science and scripture could peacefully coexist lay in how one interpreted the Bible—just as it always had.

Meanwhile, twentieth-century geologists had settled into a comfortably uniformitarian worldview. Studying processes active today, they believed, was the key to understanding the worlds of the past. Anti-catastrophist views were so embedded in conventional thinking that when a young upstart discovered evidence for an enormous flood, it took most of the century for his colleagues to accept his heretical notion. But as geologists reluctantly came to appreciate, once again, the geologic role and topographic signature of catastrophic flooding, they developed a foundation for rational explanations of many of the world's flood stories, including, some would argue, Noah's Flood.

11

The Heretic's Flood

IT IS HARD TO SEE evidence for what you're sure cannot exist. Twentieth-century geologists were no exception to this rule. They were certain that enormous floods capable of sculpting topography were impossible. Until, that is, one of them rediscovered the ability of catastrophic floods to reshape Earth's surface in the curious landscape of eastern Washington's scablands, a desolate region stripped of soil.

After teaching geology at the University of Washington for a decade, I was embarrassed that I had not yet seen the deep canyons where tremendous ice age floods scoured down into solid rock to sculpt the scablands. So when colleagues asked me to help lead a field trip there, I decided it was about time I checked out this dramatic terrain. But lead a field trip to somewhere I'd never been? No thanks, I replied, how about I just tag along? When the announcement came out, I was listed as a trip leader. Clearly this was going to be educational. The question was for whom.

Geology field trips usually involve a lot of high-speed talking in low-speed vans. As the designated distraction crossing over Washington's Cascades, I related the history of twentieth-century arguments over the timing of when the range rose to rival the Swiss Alps. Geologists working in the northern Cascades saw the range as ancient, having risen before waves of black lava flowed out from Yellowstone to cover eastern Washington fifteen to seventeen million years ago. Those working in the southern Cascades argued the range was much younger, having come up well after emplacement of the lava blanket. It turns out that there is a simple way to reconcile these fundamentally conflicting interpretations. The modern topography of the Cascade Range is a composite, the southern half rising much more recently to stand shoulder to shoulder with its elder sibling to the north. Sometimes conflict is all about perspective.

We descended the Cascades and soon entered the high desert of eastern Washington. The temperate rainforest of western Washington was miles behind us, and the lack of plants made it easy to see the landforms. Once across the Columbia River we continued eastward, driving up onto a plateau where swirling winds blew soil off freshly plowed fields. Racing the dust devils, we dropped into Moses Coulee, a canyon with vertical walls of layered basalt half buried beneath talus ramps. Nothing had removed the rocks that fell to the valley floor. They just piled up in place, right where gravity left them.

We stopped, gathered the students on a small rise, and asked them how the canyon was formed. They immediately ruled out wind and glaciers. The valley was not U-shaped like typical glacial valleys, and none of us could imagine how wind might gouge a canyon out of hard basalt. But neither did anyone see a river or stream. After a while I pointed out that we were standing on a pile of gravel and asked the class to explain how these rounded granite pebbles came to be there when the closest source of granite lay over the horizon. Silence.

Hiking through eastern Washington canyons littered with exotic boulders has long been a standard field trip for beginning geologists.

It takes a while to register what you see there: the water-scoured cliff of a now dry waterfall hundreds of feet high in the middle of the desert; giant potholes where no river flows today; granite boulders parked in a basalt canyon. Gradually, the contradictions fall into place and answer the questions of where car-sized wayward boulders came from and what was the source of the water that moved them around and carved the falls. Students can conjure up eastern Washington's giant floods once their professors give them the clues. Once you know what to look for, the evidence is hiding out in the open in plain sight. But the idea of a great flood capable of gouging deep valleys into hard rock seems unlikely in the middle of a desert, particularly when you've been taught that such a thing is impossible.

After European geologists dismissed a central role for a catastrophic flood in earth history, the idea became geological heresy. Although J Harlen Bretz uncovered evidence of giant floods in eastern Washington in the 1920s, it took most of the twentieth century for other geologists to believe him. Geologists had so thoroughly denied the existence of great floods that they could not believe it when somebody actually found evidence for one.

A controversial figure throughout his career, Bretz won no awards until long after he retired and his most influential and vociferous critics died. There was no volume written by distinguished colleagues to honor his career. He was an outsider, a heretic dismissed by the scientific establishment. A classic field geologist, Bretz figured out the story of the region's giant glacial floods, seeing what others at first could not and then would not see to sort out the pieces of a landscape-scale jigsaw puzzle.

Bretz became unpopular when he questioned orthodox uniformitarianism, Lyell's dictate that the processes of today are the same as those of the past. Fresh out of graduate school and perhaps not knowing any better, Bretz identified compelling evidence for a gigantic flood. A reluctant heretic, he insisted on valuing field evidence above theory, piecing together the story of how a raging wall of water hundreds of feet high roared across eastern Washington, carving deep channels before cascading down the Columbia River gorge as a wall of water high enough to turn Oregon's Willamette Valley into a vast backwater lake. This time it was the scientific community that refused to see the evidence. Vying to be the first to prove himself wrong, Bretz kept digging. But as he kept finding more evidence of a really big flood, the geological establishment kept coming up with ways of explaining it away.

Bretz taught in his native Michigan before heading west to teach high school in Seattle. A field enthusiast, he spent his weekends and summers studying the geology around Puget Sound as well as glaciers in the nearby Cascade Range. Eventually he enrolled at the University of Chicago, graduating *summa cum laude* with a PhD based on western Washington's glacial geology in 1913. After spending a year on the faculty of the University of Washington, where his colleagues did not appreciate his enthusiasm for fieldwork, he accepted an invitation to return to Chicago, where he taught until he retired in 1947. Dedicated to teaching geology in the field and enamored with the landscapes of eastern Washington, he started bringing summer classes to the Columbia River gorge.

There Bretz found exotic granite boulders perched on basalt cliffs hundreds of feet above the highest recorded river level. Glaciers could not have carried these boulders to these elevations. Geological evidence had already proven glaciers had never reached the gorge. His colleagues thought this part of the Cascade Range lay submerged beneath the Pacific Ocean when the boulders arrived, carried by floating ice. Finding no evidence of marine fossils or ancient beaches, however, Bretz concluded the boulders must have been deposited by fresh water. But what could have generated such an enormous flood?

Each summer he returned to explore farther upstream. After several summers canvassing the gorge, he shifted north to the scablands.

Exploring the strange topography of the area, Bretz came across dry waterfalls and potholes hundreds of feet above the modern river. Gigantic gravel bars deposited within coulees (dry valleys) implied deep, fast-flowing water. Scour lines that crossed over drainage divides showed that flowing water had overtopped ridgelines and spilled into adjacent valleys. Streamlined hills rose like islands sticking up more than a hundred feet above the scoured out channelways. Bretz realized the chaotic landscape had been carved by enormous floods that chewed deep channels through hundreds of feet of solid basalt. Here, right before his eyes, was the unthinkable.

Ever since Reverend Samuel Parker first described the Grand Coulee as a former channel of the Columbia River in 1838, explorers and geologists agreed that a glacially diverted river that ran across the plateau gradually carved the scablands before returning to its normal valley. But Bretz identified how these now streamless canyons defined a drainage pattern unlike any formed by normal rivers. Here was an interconnected complex of enormous channels that branched out only to reconnect downstream. Such a network could only form if water had filled valleys to overflowing and spilled a great flood over their drainage divides. He called this enormous flood the Spokane Flood. But what was the source of all that water?

Bretz first presented his thoughts on the channeled scablands to the Geological Society of America in 1923. Focused on describing his field observations, he was careful not to invoke the taboo of referring to a monstrous flood. He attributed the flows that carved the valleys to an ice dam across the Columbia River that forced water to spill out across the scablands. Over successive summers Bretz became increasingly confident that the scablands were not just the gradually produced work of a diverted river.

He recognized that 100-foot-high piles of gravel on the canyon floors were built by even deeper flows and that the hanging valleys that drained over dry waterfalls were not the product of normal stream erosion. These features were carved by a process that shut off before forming a smoothly integrated channel network. Troubled by how the field evidence pointed to a giant flood spilling over from the Columbia River to scour the scablands, he found good reasons to reject all other possibilities.

Tracing the evidence downstream through the Columbia River Gorge, Bretz found that his flood deposited an enormous delta around Portland, Oregon, backing up flow into the Willamette Valley. Taking advantage of locations where the flow constricted, he calculated a peak discharge so large he even doubted it himself—over sixty-six million cubic feet per second. Field evidence kept pointing to a really big flood.

Bretz could think of only two forces that could have produced his troubling flood. Either a very rapid and short-lived warming of the climate or a volcanic eruption beneath an ice cap. Neither provided a satisfying explanation. He had a flood without a cause.

His colleagues were as perplexed as he was. The battle over how long it took rivers to carve valleys had already been won. Even Bretz acknowledged the challenge that his catastrophic flood presented to conventional thinking. Yet, the giant gravel bars in the now dry canyons mirrored the form of ripples on the bed of a sandy river—only they were much, much larger. Some tear-dropped shaped bedrock hills were still capped with loose silt, showing that the flow that streamlined them did not overtop them. One could map the extent of the flood. Somewhat reluctantly, Bretz concluded that catastrophic flooding provided the best explanation for his field observations.

In January 1927, Bretz was invited to present his findings to the Geological Society of Washington, DC. It made no difference that he systematically outlined his arguments: dry canyons carved hundreds of feet into hard basalt, hundreds of dry waterfalls some two to three miles across; the stripping off of several hundred feet of silt and soil over large areas; and interconnected overflow channels that crossed drainage divides. It was an ambush. Representatives of the Geological Society of America and the U.S. Geological Survey also attended the meeting. One by one they rose to crucify the heretic's description of the flood.

The first critic cautiously warned about the difficulty in finding a source that could release so much water so fast. Surely, he asserted, many small floods gradually carved the scablands. The next critic doubted that so much hard basalt could be carved out in a brief flood, no matter how deep the flow. Another argued that a diverted Columbia River swollen by glacial meltwater could have slowly incised the scabland channels. This defender of geologic sanity was so eager to dismiss the idea of a catastrophic flood that he argued that the elevations of spillways originally cut at different altitudes were now identical due to subsequent earth movements fortuitously aligning them at precisely the same height. Although no one questioned his observations, every speaker challenged Bretz's interpretation, pointing out that he had no way to explain how to get so much water so fast. Although this was a lopsided debate, it was deeply rooted in a long tradition of geologists sparring and arguing over how to interpret observations. What everyone, including Bretz, could agree on was that further fieldwork was needed to explain the channeled scablands.

One of those attending the DC ambush was Joe Pardee, a Geological Survey geologist. Two years before, Pardee wondered whether the scablands could have been carved by catastrophic drainage from Lake Missoula, an ancient glacier-dammed lake he had discovered evidence for in western Montana back in 1910. He wrote to Bretz suggesting this as a possible source for his Spokane Flood. Bretz ignored him. An account of the DC meeting relates how during the discussion Pardee confided to a colleague that he knew the origin of Bretz's flood. But with a career and a reputation to maintain he stayed quiet, unwilling to upset his boss, who had been the first speaker to challenge Bretz.

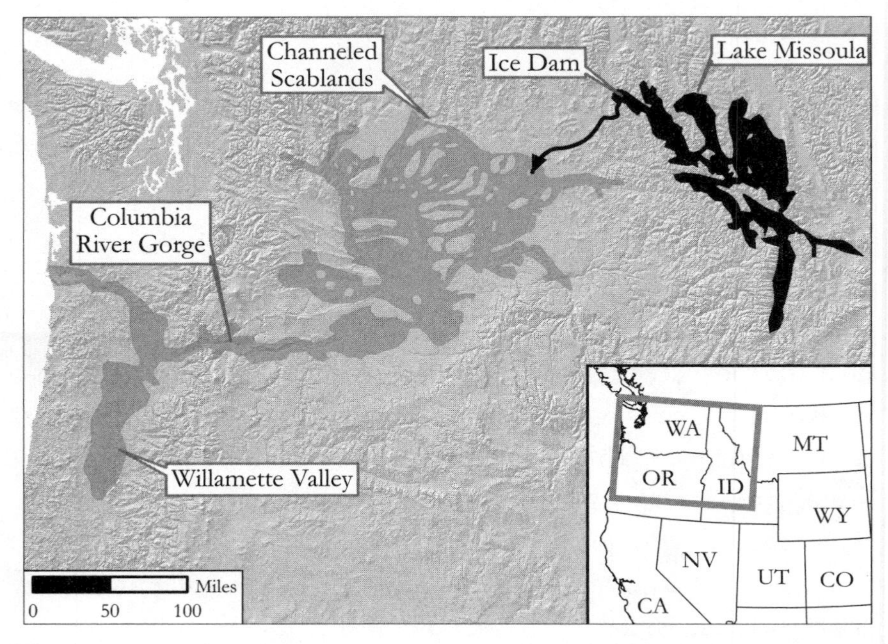

Map of floods from ice-dammed glacial Lake Missoula (black) showing branching structure of flood through the Channeled Scablands and the extent of backwater up into the Willamette Valley (gray).

The next summer, Bretz went back to the field and found deposits formed by water flowing backwards up Columbia River tributaries. Along the Snake River he traced backwater deposits upstream beyond Lewiston, Idaho. Only a huge flood could have sent deep water surging back up into tributaries. Grudgingly taking Pardee's hint, Bretz settled on drainage from Lake Missoula as the source of his flood.

Nobody else believed him. The remoteness of the scablands and the difficulty of traveling there during the Depression fostered skepticism among colleagues back east, few of whom had been out west to see the area for themselves.

Geologists kept arguing about the scablands and attacking the heretic's flood. One colleague dusted off Bretz's original idea of a flood produced by water backing up from a Columbia River ice jam. Others invoked glacial erosion, ignoring the fact that the scablands lay south of the well-documented extent of glaciers. Or maybe ancient rivers slowly cut down through normal river erosion, although this could not account for the giant potholes. One prominent critic carefully described—and then ignored—areas of ten-foot-high undulating ridges running across the trend of the valley they occurred in, features Bretz later identified as incontrovertible evidence of catastrophic flooding.

The tide turned at the 1940 Seattle meeting of the American Association for the Advancement of Science. In a session on the glacial geology of the Pacific Northwest crowded with megaflood skeptics, Joe Pardee described evidence for giant ripple marks on the bed of the glacial-age Lake Missoula. Pardee recognized that the fifty-foot-high ripples on the lakebed were formed by fast-flowing currents rather than the sluggish bottom water of an impounded lake. Only sudden failure of the glacial dam could have released the two-thousand-foot-deep lake. Pardee did not need to point out that here was a logical source for Bretz's flood: the catastrophic release of six hundred cubic miles of water through a narrow gap would sweep away everything in its path.

In 1952 Bretz returned for a last summer of fieldwork in the scablands. Nearly seventy years old, he wanted to see evidence uncovered by the Bureau of Reclamation's Columbia Basin project. He was delighted to find their excavations showed that the hills he interpreted as hundred-foot-high gravel bars were indeed formed by deep, fast-flowing water.

Examining the bureau's aerial photography, Bretz found the smoking gun that clinched his story. The bird's-eye view revealed the rugged rise and fall of the topography he recalled scrambling over decades before to be giant ripples like those hundreds of miles upstream at the outlet to Lake Missoula. Hidden beneath the sagebrush, the field of megaripples was strikingly obvious from the air.

There was "no other explanation for their rhythmic patterns than that of bedform development by amazingly deep, swift flood water."[1] Bretz had been right all along.

It had taken decades, but he finally had the evidence to convince skeptical colleagues. In August 1965, an international delegation of geologists traveled from Lake Missoula down through the scablands to see the evidence firsthand. Bretz was no longer able to travel, so at the end of the trip the delegation sent him a congratulatory telegram that ended with "We are now all catastrophists."[2] It took a changing of the generational guard for geologists to accept the heretic's flood.

In the summer of 1976, just before his ninety-fourth birthday, NASA scientists invoked Bretz's careful fieldwork on understanding features diagnostic of catastrophic flooding to explain streamlined hillslopes and giant channels in images returned by the Viking spacecraft orbiting Mars. Half a century after government scientists gathered to denounce his radical theory of erosion by catastrophic flooding, NASA was hailing his studies of the channeled scablands as the key to understanding enigmatic Martian landforms.

In 1979, the Geological Society of America awarded Bretz its highest honor, the Penrose Medal. He was ninety-seven years old and is reported to have jokingly complained to his son, "All my enemies are dead, I have no one to gloat over." In hindsight, he described his work as a struggle against the dominance of uniformitarian thinking that prejudiced his colleagues against the idea of a great flood:

> *Was not this debacle that had been deduced from the Channeled Scabland simply a return, a retreat to catastrophism, to the dark ages of geology? It could not, it must not be tolerated. . . . They demanded, in effect, a return to sanity and Uniformitarianism.*[3]

Later fieldwork by others revealed evidence for many floods, each of which left a single thin layer in thick stacks of backwater sediments. The ice dam had failed over and over again. When an advancing glacier dams a river, it can prove stable until the water backs up deep enough to float the ice, catastrophically undermining the dam. Once the lake drains, the ice can readvance, repeating the whole process again and again until the glacier finally retreats. The ice dam blocking Lake Missoula's only outlet had become a virtual flood machine.

Calculations accounting for the estimated rate of flow into the lake indicated that it took three to seven decades to fill, the same time interval between lake-draining floods revealed by the number of annual sediment layers in the lake bottom sediments. Downstream, careful stratigraphic analyses showed that each layer of flood-deposited sediment represented a separate event with velocities exceeding twenty feet per second. Radiocarbon dating of organic matter deposited in the flood sediments revealed there were as many as 100 separate floods as the ice dam formed, failed, and reformed every few decades from 15,300 to 12,700 years ago.

In a way, the finding that Lake Missoula failed scores of times brought Bretz's heretical idea back into line with uniformitarian thinking. Glacial dam failure is a simple process to understand. It works via the mechanics of floating an ice dam. Fill up a glacially dammed lake enough to float the dam and, presto, you get an instant catastrophe. Keep filling it up and you get a repeating series of catastrophes.

Recognition of the Missoula Floods helped identify similar landforms in Asia, Europe, Alaska, and the American Midwest, as well as on Mars. There is now compelling evidence for many gigantic ancient floods where glacial ice dams failed time and again on the margins of great ice sheets. In hindsight, it's obvious that ice dams are not all that intelligently designed for the simplest of reasons—they float.

At the end of the last glaciation, giant ice-dammed lakes along glacial margins in Eurasia and North America repeatedly produced catastrophic outburst floods. Ice dammed north-flowing Siberian rivers, spilling them over drainage divides and changing their courses. England's destiny as an island was sealed by erosion from glacial outburst floods that carved the English Channel. Devastating floods were a fact of life on the margins of the world's great ice sheets.

We now know that large ice dam failures were common in prehistoric North America and Eurasia. And since ice dams tend to fail catastrophically, people living around ice sheet margins probably witnessed giant floods. Could survivors of such events have passed their stories down through the ages?

A campsite with charred bones and stone artifacts buried under pre-flood deposits along with a stone artifact recovered from a giant flood-deposited gravel bar along the Columbia River provide the only reported physical evidence I could find that anyone could have witnessed a thousand-foot-high wall of water crashing through the Columbia River gorge—and no indication of whether or not any possible human witness lived to tell about it. Early missionaries in eastern Washington reported that Yakama and Spokane Indians had oral traditions of a great flood that described locations where survivors sought refuge. The native inhabitants of the lower Columbia River also reportedly had a legend of a catastrophic flood. Upstream in Idaho, the Nez Perce and Shoshone also had flood stories. Downstream, the Santiam Kalapuya people of the southern Willamette Valley, Oregon, had a story of a time the valley filled with water, forcing all the people to flee up a mountainside west of Corvallis before the waters receded.

One problem with attributing such stories to the Lake Missoula floods has been that the floods occurred before the generally accepted time of human arrival in North America. However, the recent discovery of human coprolites (fossilized excrement) radiocarbon dated to 14,000 to 14,270 years ago at Paisley Caves, in south-central Oregon, places human populations in the region during the time of the Missoula Floods. If the region's flood stories do record the Missoula Flood and backwater flooding of the Willamette Valley, then it means that science is only now catching up with folklore.

In North America, glacial dam failures were not restricted to the Pacific Northwest. Catastrophic drainage of glacial Lake Agassiz, a vast lake that formed in a moatlike depression on the edge of the retreating Canadian ice sheet, happened numerous times as the lake's shoreline kept shifting as the ice melted off. Exposure of new outlets sent great floods cascading off in different directions, south to the Mississippi, north through Hudson's Bay, and east down the St. Lawrence River to the Atlantic. When the lake finally emptied for the last time more than eight thousand years ago, it was still a hundred times larger than Lake Missoula, releasing a pulse of freshwater big enough to change ocean circulation and shut down the current of the Gulf Stream, which brings warm water to the North Atlantic and keeps northern Europe habitable (without it Britain would have a climate like Siberia's). It's no coincidence that cold periods recorded in Greenland ice cores correspond to major drainage events from Lake Agassiz.

Given that the ancestors of Native Americans from Alaska to Tierra del Fuego are thought to have come from Asia via the Bering Strait (whether overland or by paddling along the coast), they would have passed near the ice sheet margin. The Native American Clovis culture overlaps with the great outburst floods in the Midwest and Northeast that continued to occur until glacial Lake Agassiz drained for the last time around 8,400 years ago. Algonquin flood stories center around the Great Lakes region, along the shifting outlets of ice-dammed Lake Agassiz. Downstream, in Nebraska and Kansas, Pawnee stories associate the bones of giant bison with catastrophic floods along the Missouri River. Do these stories relate

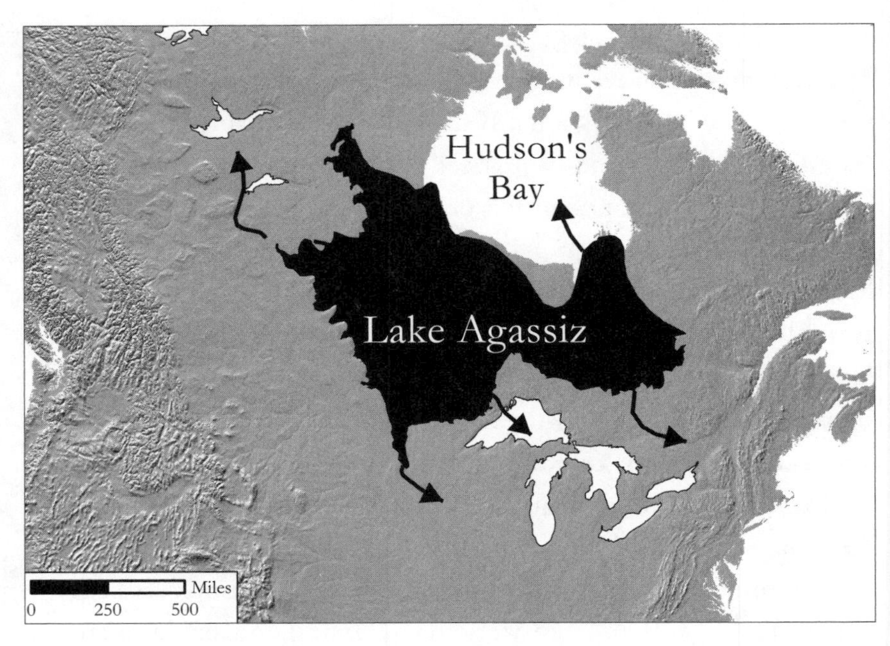

Map of glacial Lake Agassiz showing its maximum extent and outlet directions for various mega-floods during deglaciation.

ancestral tales of dramatic disasters, or attempt to explain puzzling features of the local environment, or both?

An intriguing Ojibwa (Chippewa) legend from around Lake Superior tells of a devastating flood at the beginning of time when a great snow fell one September. A bag contained the sun's heat until a mouse nibbled a hole in it. Spilled warmth instantly melted all the snow, producing a huge flood that rose above the tops of the highest pines. Everyone drowned except for an old man who drifted about in his canoe rescuing animals. It doesn't take much imagination to see this as the story of an ice dam failure.

Stories about ice dam failures also come from Northern Europe. Nordic mythology tells of how a kingdom of ice, ruled by an ice giant, once covered Scandinavia. When the Norse god Odin and his brothers killed the frozen king, his blood (water) gushed forth to drown the other ice giants. In one story, Odin and his siblings used the frigid giant's eyebrows to make a wall separating the land of ice from the land of people. This boundary sounds suspiciously like the snaking ridges of glacial debris (called moraines) left by ice retreating across Sweden and Finland. Viking songs and stories written down and preserved in Iceland before 1250 AD also tell of how the modern world began when Odin and his brothers slew an ice giant, releasing a great flood that inundated the lowlands and drowned large mammals.

Flood stories from tropical climates have different narrative details. Accounts of big floods from throughout the Pacific Islands describe rapid inundation as a huge wave from the sea tears up trees and forces survivors to high ground. Many South Pacific flood stories fail to mention rain at all. Instead, in these accounts, the sea rose to flood all but the highest places. The remarkable tsunami stories from Sumatra, Borneo, New Guinea, Fiji, Tahiti, Tonga, New Zealand, and Hawaii show how tales of infrequent local disasters can become the stuff of legends.

Tsunamis come out of the blue, from over the horizon—sometimes from across an entire ocean. When a big shock like a landslide or an earthquake displaces a lot of water, the pressure wave travels. The surrounding ocean water doesn't. Moving at tremendous speed, the resulting wave can cross an ocean in a few hours. Typically, the leading part of the wave arrives as a water-level depression when it grounds out as it approaches shore. So the water rushes out and then surges back in as the crest of the wave arrives. All too often, the initial mystery of the falling tide and the seductive exposure of bare seabed attract the curious before a surging wall of water sweeps away everyone in its path. With no local cause to invoke, divine displeasure might seem like the only reasonable way to explain monstrous rogue waves. Fijians are said to have only recently stopped keeping great canoes ready in case of a surprise flood from the sea.

The Indian Ocean tsunami resulting from the December 26, 2004, magnitude 9.3 earthquake killed more than a quarter-million people. Hard hit by the tsunami, Simeulue Island in Indonesia's Aceh province lost only seven people out of a population of almost eighty thousand. What explains such a low casualty rate? The islanders had an oral history recounting another massive tsunami that struck in 1907, killing three-quarters of the island's inhabitants and stranding bodies in the tops of coconut trees. Survivors of the 1907 disaster made up a new word for "the ocean coming onto the land." Interviews after the 2004 tsunami revealed that the story did its job. When the ground shook the locals knew to flee their low-lying coastal villages and head for the hills. There were numerous casualties on the mainland, where the population had no such oral history of a previous tsunami.

Native American stories of a flood coming from the sea are common along the Cascadia subduction zone, from Northern California to the Oregon and Washington coasts and north to Vancouver Island. Tremendous earthquakes shake this region each time the oceanic crust beneath the Pacific gets shoved a little farther under North America. We know the last major subduction zone earthquake occurred on January 26, 1700 because Japanese temple records tell of a mysterious tsunami arriving without any ground shaking. The wave generated on the west coast of North America traveled all the way to Japan.

Early accounts from the Pacific Northwest record that flood traditions were common among coastal tribes. Missionaries were puzzled that some stories recounted floods just three or four generations back. One old man of the Clallam tribe said his grandfather had even met a survivor of the great flood. Missionaries wondered how native peoples could have been so confused about the timing of Noah's Flood. They weren't. Such stories read like eyewitness accounts because tsunamis devastated their ancestors' coastal communities. Archaeological evidence documents that villages along the British Columbia, Washington, and Oregon coasts were inundated by tsunamis and abandoned after the 1700 earthquake. After the ground shook violently for more than three minutes, a thirty-foot wave smashed into the coast. The dramatic tale was sure to be retold by survivors.

Older traditional stories from throughout the region tell of ancient struggles between Thunderbird and Whale, graphically describing ground shaking and accompanying flooding from the sea. These stories depict Whale as a monster terrorizing animals and depriving people of food. Seeing that the people were starving, benevolent Thunderbird flew from his mountain home and dove into the ocean to battle Whale. During their struggle the sea fell and rose again, sending canoes into treetops and killing many people.

Even Western mythology has direct links to tsunami stories. An unusual Mediterranean tsunami may explain both the Greek story of Deucalion's flood and the myth of Atlantis, the fabled city that sank into the sea. In 1960, Greek seismologist Angelos Galanopoulos proposed that the volcanic destruction of the island of Santorini (also known as Thera) was responsible for the story of Deucalion's flood. Radiocarbon dating of the ash from the eruption of Santorini (as the volcano composing the island was also known) revealed it dated from 1500 to 1600 BC, around the historical reign of King Deucalion. On the island of Paros, a marble pillar listing the kings of Greece implies Deucalion's flood occurred in about 1539 BC. The eruption destroyed a great city on Santorini and generated a tsunami that ravaged the Greek coast. In an early version of the Deucalion story the flood is even said to have come from the sea.

The connection to the story of Atlantis comes through Plato, who believed it had been handed down since the time of the great lawgiver Solon. Two hundred years before Plato's time, Solon traveled to Egypt and asked priests there about Deucalion's flood. They told him of a great disaster that had destroyed the mighty island metropolis of Atlantis. Lying at the center of three concentric harbors connected to the sea by a narrow channel, the great city vanished in a single day. This island kingdom beyond the Pillars of Hercules, which Plato placed past the Straits of Gibraltar, existed nine thousand years before Solon's time.

Galanopoulos suggested that Solon had mistranslated the Egyptian word for one hundred as one thousand, because when divided by ten, Plato's age for Atlantis comes out at about 1500 BC and its size matches that of Santorini. Did Plato realize that Solon's oversized island would not fit in the Mediterranean, and did he move the Pillars of Hercules from the southern Peloponnesus to Gibraltar, expediently banishing Atlantis to the unexplored world beyond?

Whether or not they lived in Atlantis, the inhabitants of Santorini built their city on the flanks of an active volcano. They chose the easily defended island because it was ringed by the natural moat of a volcanic caldera pleasantly plumbed with geothermal hot water. In exchange for the Bronze Age luxury of running hot water, residents unwittingly took on the risk of living in a city that lay within the heart of an active volcano. Eventually, the catastrophic eruption of their island home obliterated their idyllic city and triggered a tsunami. I suspect that this event is immortalized in the story of Deucalion's flood.

Others have argued that a more gradually rising sea level, and not a tsunami, was responsible for the world's flood stories. In 1960, Rhodes Fairbridge, a Columbia University geology professor, proposed that flooding of coastal lowlands around the world displaced human communities and spawned ancient flood stories when sea level rose by several hundred feet as the ice caps melted at the end of the last glaciation. His hypothesis did not win many converts, as no one would need an ark to escape a sea rising less than a foot a year.

However, the collapse of the North American ice sheet during

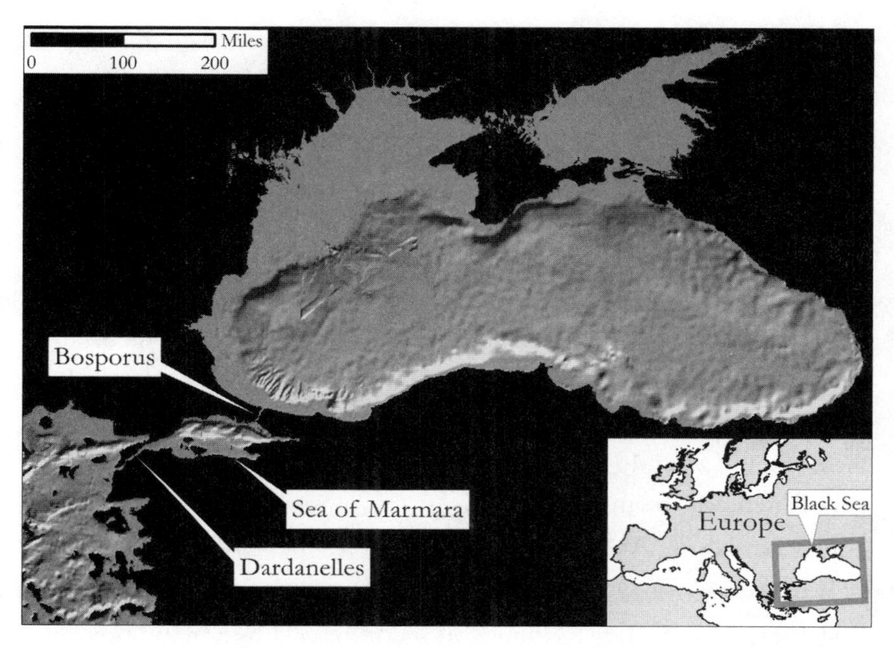

Map of the Black Sea showing connection to the Mediterranean through the Dardanelles, Sea of Marmara, and Bosporus.

deglaciation did cause a rapid five-foot rise in sea level between 8,300 and 8,200 years ago, right about the time of the last big flood from Lake Agassiz. The rising sea flooded coastal areas across Europe and led to the sudden loss of land favored by early farmers. This event coincided with an abrupt migration of Neolithic peoples and an expansion of early agriculture into areas previously occupied by hunter-gatherers. Earlier, Neolithic sites in Europe were restricted to Anatolia and Greece. Following the abrupt rise in sea level, farming began spreading across the continent. How did a rising sea level trigger a Stone Age migration?

It has been suggested that this jump in sea level catastrophically breached the low ridge of the Bosporus, spilling the Mediterranean into a low-lying freshwater lake in the Black Sea basin. This event would have submerged almost 28,000 square miles and topped the lake up to sea level. This may have submerged some of the earliest farming communities, sending refuges off in all directions.

In the spring of 1969 the oceanographic research vessel *Atlantis II* found a remarkable layer of organic matter in the bed of the Black Sea. Sandwiched as the middle of three distinct sedimentary layers, the black mud recorded how a former sea turned into a freshwater lake and then back into a sea. In some places half of the curious black mud was composed of plant and animal remains. The organic muck lay atop unusual gray clay with fresh water in its pores. The saltwater fauna of the lowest layer was replaced by freshwater organisms, which were then replaced by saltwater species sometime later. Apparently, the Black Sea had been a freshwater lake when the sea level was lower and rivers were swollen with glacial meltwater. Then an influx of seawater shifted the bottom of the water body from well oxygenated to stagnant, oxygen-poor conditions. When did this happen? Radiocarbon dating of the organic matter in the strange layer of black mud indicated that the rush of seawater occurred about seven thousand years ago.

In 1972 the Victoria Institute, a Christian society established in 1865 with the professed mission to reconcile apparent discrepancies between the latest geological findings and scripture, held a symposium on Noah's Flood. There, British Bible-science enthusiast Robert Clark suggested that the biblical flood deposited the organic-rich mud at the bottom of the Black Sea. Perhaps the sea level rose enough to spill into the Black Sea when a large piece of Antarctic ice calved into the sea, or when a volcano erupted beneath the ice cap. However it happened, Clark thought the stagnant conditions at the bottom of the Black Sea ensured preservation of a flooded landscape deep below the surface. Few took seriously his suggestion that Noah's hometown might lay entombed beneath the mud of the Black Sea.

Yet since antiquity we've known water flows both ways between the Mediterranean and Black seas. The lighter, fresher water of the

Black Sea flows out above a reverse current of denser salt water that flows from the Mediterranean along the bottom. Up until the invention of steam power, mariners traveling upstream to the Black Sea pulled themselves through the Dardanelles and Bosporus by lowering baskets full of stones down into the strong bottom current, which then dragged their boats against the surface current. Hugging the seafloor, a submerged river of salt water flowed into the Black Sea.

In 1993, oceanographers Bill Ryan and Walter Pitman led a joint Russian-American expedition to survey and sample the floor of the strategically important Black Sea. Scanning the seabed with sonar, their team found evidence of ancient streambeds, river-cut canyons, and submerged shorelines. In samples of the bottom sediments saltwater mussels replaced freshwater mussels at the transition from the gray clay to the strange black mud above it. When their carbon dates came back from the lab they were astounded to find that the first marine creatures that invaded the freshwater lake were the same age no matter where and at what depth they sampled. Oxygen depletion and saltwater intrusion started simultaneously throughout the Black Sea, exactly what one would expect if a sudden flood of salt water smothered a great freshwater lake.

High-resolution profiles of subsurface layers, mapped by setting off small explosions and measuring the travel time of the resulting seismic waves, revealed a former land surface buried in the seafloor sediments. The unconformity defined by the contact between the layers of sediment above and below this surface extended to depths well below the bedrock sill at the Bosporus. Drill cores punched into and brought up from the seafloor contained subaerial desiccation cracks and in-place roots of shrubs covered by marine mud. Changes in the isotopic composition of different layers in the cores showed it took about a thousand years for enough seawater to pour into the Black Sea basin from the Mediterranean to begin supporting marine life on the seabed. A later expedition in 2000 discovered evidence for a shoreline with a cobble beach hundreds of feet below the modern waves.

Ryan and Pitman knew that farming had been practiced in the region for at least a thousand years before the Mediterranean spilled into the Black Sea. Anyone living in the fertile valley would have been forced to flee with their livestock as their world disappeared beneath the rising waters. Archaeologists had found that this time coincided with the onset of the initial migration of farming cultures into Europe and the floodplains of Mesopotamia. Here was another candidate for a reasonable explanation of Noah's Flood.

Other scientists have challenged Ryan and Pitman's interpretation of a sudden influx of salt water into the Black Sea. The assemblage of microscopic marine creatures (foraminifera) recovered from cores drilled into the bed of the Marmara Sea, the water body that connects the Black Sea to the Mediterranean, suggests an earlier, less catastrophic reconnection, and a more gradual flooding than that inferred by Ryan and Pitman. In addition, the elevation of delta deposits at the pre-flood mouth of the Danube River, where it drained into the Black Sea, constrains the pre-flood water level in the Black Sea to having been less than a hundred feet below modern sea level. This means that Ryan and Pitman's flood could have raised the water level in the Black Sea by no more than that amount. While the geological community was divided over the Black Sea flood hypothesis, most of the authors in a 2007 volume dedicated to examining its geological basis argued in favor of a gradual, noncatastrophic reconnection of the Mediterranean and Black Sea over the past 12,000 years. Lively controversy characterizes ongoing geological debate over the so-called Noah's Flood hypothesis.

When I first heard Ryan and Pitman's theory, back in the 1990s, it made sense to me. It sounded like a reasonable explanation for the story of Noah's Flood. However, at the time I didn't know about the Sumerian tablets identifying Ziusudra as the last king of Shurrupak before a Mesopotamian flood. I now believe that there is no way to tell whether Noah's Flood was the Black Sea flood or a major Mesopotamian flood. No matter how intriguing either idea may sound, both offer seemingly reasonable explanations.

Wherever they came from, the first farmers arrived in southern Mesopotamia shortly after the filling of the Black Sea. Sumerian cities sitting on the undisturbed ruins of these first farming towns without any archaeological evidence of distinct breaks in culture suggest that these early farmers were the ancestors of the Sumerians. Did they bring the story of a great flood that destroyed their world with them when they fled to Mesopotamia from an ancestral homeland now at the bottom of the Black Sea? If so, periodic flooding would have reinforced the tradition of a great flood among those living on the low ground between the Tigris and Euphrates rivers.

Creationists quickly denounced Ryan and Pitman's claim of scientific support of the biblical flood. This was not their global deluge. Grand as it was, the Black Sea flood could not be Noah's Flood; it was still too puny. An influential creationist website even accused Ryan and Pitman of trying to destroy the Bible. Other creationists simply maintained that Satan had clouded the minds of those denying the reality of a global flood.

There was a time when both geologists and conservative Christians would have interpreted the evidence of a catastrophic Black Sea flood as proof of Noah's Flood and confirmation of the historical veracity of Genesis. But times have changed. Now geologists present evidence in support of Noah's Flood, and creationists hold out for belief in a global flood for which no evidence can be found. Yet, who's to say that the original Noah wasn't among those living in the area now submerged beneath the Black Sea? At this point an answer lies beyond the reach of geological, archaeological or historical inquiry. To those with opinions about such things, the truth remains a matter of faith.

12

Phantom Deluge

I HAVE OFTEN WONDERED how creationists could reject a whole series of independent scientific advances, from the coherent order of species in the fossil record to radiometric dating and plate tectonics. Even more curious is how they reject science even when it appears to support the historical veracity of scripture. The key to understanding modern creationist thinking lies in understanding the influence of John Whitcomb and Henry Morris. Shocked by what they saw as the bending of God's Word to the whims of science, these two men wrote the book that launched the modern revival of young-Earth creationism.

In 1948, Whitcomb was a Princeton ancient and European history major who converted to evangelical Christianity in his freshman year. Following graduation, Whitcomb enrolled in Grace Theological Seminary, a fundamentalist school in Winona Lake, Indiana, where he later taught the Old Testament and Hebrew. This combative young Bible teacher, the son of General Patton's chief of staff, considered endorsement of an old Earth and a local Flood to be an abominable folly rooted in uncritical acceptance of uniformitarian geology.

Raised Southern Baptist, Morris drifted into religious indifference in his undergraduate years. After a period of intense soul-searching following graduation, he rejected evolution and embraced a literal six-day creation. Not letting this interfere with pursuing his worldly interests, he went on to earn a PhD in hydraulic engineering from the University of Minnesota. A successful academic, he eventually headed the civil engineering program at Virginia Polytechnic Institute.

This unique pair, an Old Testament theologian and a hydraulic engineer, met in the summer of 1953 at the annual American Scientific Affiliation convention. Whitcomb attended Morris's presentation on "The Biblical Evidence for a Recent Creation and Universal Deluge." He was as impressed with the talk as he was appalled by its polite dismissal by an audience familiar with J. Laurence Kulp's devastating critiques of flood geology. Whitcomb had found an ally.

Further incensed by the favorable reception of Bernard Ramm's book in the evangelical community, Whitcomb decided to write his thesis on the biblical case for a global flood. He completed his dissertation in 1957 and immediately began looking for a publisher. Two established evangelical publishers, Eerdmans and Moody, expressed interest. After seeing the manuscript Eerdmans declined to publish it. Moody agreed to take it, but encouraged Whitcomb to have the chapters dealing with scientific aspects of the Flood either checked or coauthored by a PhD scientist, preferably a geologist. Whitcomb reluctantly agreed.

The only geologist he could find willing to look at the manuscript was appalled by what he read. He wrote to Whitcomb that if there were any truth to such a globe-wrecking flood, some well-trained geologist would have put the story together. The reviewer suggested that Whitcomb learn the basics of historical geology.

Instead, Whitcomb decided to limit himself to advice from fellow creationists. Of those he approached, Morris turned out to be the most helpful and enthused. Impressed by the first three chapters of Whitcomb's dissertation, Morris admitted that he, too, had been working on a flood geology book. He advised Whitcomb to refrain from sarcasm and ridicule and suggested he emphasize theological arguments. In this way, Whitcomb could avoid getting trapped by geological ones.

Grateful for input from someone more familiar with the technical objections to flood geology, Whitcomb asked Morris to coauthor his book. Morris enthusiastically signed on. Four years later, Moody turned down their finished manuscript, but a small publisher eager to challenge the geological foundation for evolution published *The Genesis Flood* in 1961.

The book began with a long argument for the reality of a global flood that killed off everyone and everything not aboard the ark. Whitcomb and Morris acknowledged more of the recent archaeological and geological evidence than had George McCready Price, but they were just as selective and prejudicial in evaluating that evidence. They were forthright in admitting as much: "We take this revealed framework of history as our basic datum, and then try to see how all the pertinent data can be understood in this context. . . . It is not a scientific decision at all, but a spiritual one."[1]

In their view, Christians faced a stark choice: "Either the Biblical record of the Flood is false and must be rejected or else the system of historical geology which has seemed to discredit it is wrong and must be changed."[2]

Confident God's Word could not lead them astray, Whitcomb and Morris were clear about how to reconcile science and the Bible. Rejecting the idea of reinterpreting scripture to accommodate science, they advocated "letting the Bible speak for itself and then trying to understand the geological data in the light of its teachings."[3] In other words, they tried to figure out geologic history by reading the Bible and then looked for data supporting the proper conclusion—and dismissed or ignored contrary evidence.

In setting up their argument, Whitcomb and Morris first asserted biblical inerrancy and rejected both a tranquil and a local flood as inconsistent with the plain meaning of the biblical story. Any fool could see the Flood was violent and global.

Whitcomb and Morris offered a number of geological inferences from the biblical account. They were certain that a tremendous quantity of water poured down on the earth in a torrential downpour that continued for forty days and nights all around the world. Yet they also accepted that the clouds held nowhere near enough water to trigger a global flood. The floodwaters had to come from somewhere else. Pockets of water trapped underground since the Creation must have erupted to the surface. Still, this wasn't enough. They looked to the heavens for more.

They found enough water for Noah's Flood in the cryptic biblical reference to the waters above the firmament (Genesis 1:7), arguing that God enclosed the primordial world in a gigantic canopy of water vapor (the same argument astronomer Edmund Halley used in the seventeenth century). At a loss to explain rationally how to bring their vapor canopy down to earth, they invoked another miracle to collapse this curtain of water. God hung it above the sky, so He could drop it when he pleased.

Shielding the planet from the harmful effects of radiation, this vaporous cocoon allowed Adam and the antediluvian patriarchs to live incredibly long lives. The greenhouse effect it produced warmed the planet into a tropical state, explaining why fossils the world over seemed to have lived in warmer times.

However God did it, the water from the canopy spilled out over forty days and nights, draining into collapsed lowlands that became the world's oceans. Tectonic movements raised continents as the chaotic flood buried fossils in sediments that then solidified to become rocks. Then for months afterwards the world convulsed with earthquakes and volcanic eruptions, followed by a brief ice age. Whitcomb and Morris went on to claim that the rock record supported all of these inferences. Their evidence: "Almost all of the sedimentary rocks of the earth, which are the ones containing fossils and from which the supposed geologic history of the earth has been largely deduced, have been laid down by moving waters."[4]

Their technical argument for a global flood was that sedimentary rocks exist and are deposited by flowing water. From this foundation they leapt to the conclusion that a global flood, Noah's Flood, did in fact occur.

In attacking conventional views, they quoted a geological textbook out of context to argue that geologists use fossils to determine the relative age of rocks, overlooking how stratigraphic order was established unambiguously in places like Siccar Point and the Grand Canyon. Aware that geologists viewed Price as a crackpot, Whitcomb and Morris nonetheless adopted his idea that the order to the fossil record actually recorded different environments in different parts of the pre-Flood world. They offered three ways to explain why the oldest rocks contain only single-celled creatures and why younger rocks contain progressively more diverse and complex organisms. Their first suggestion was Woodward's long-discredited idea about sediment and fossils settling out by density. Their second was that marine fauna would have perished first and therefore be interred in deeper strata. Lastly, certain animals, whether by anatomical design or ingenuity, struggled longer to resist the Flood, their bodies settling later into higher layers of flood-deposited sediment.

Anyone in an introductory geology course could readily address how these ideas are incapable of explaining the fossil record. Most damning is the remarkable order to fossil sequences. Trilobites only occur in the lowest strata, which do not contain the densest fossils and often host delicate floating creatures. Were hydraulic sorting to explain the order to the fossil record, small trilobites would always be found above larger trilobites because objects of the same density sort by size when settling through a fluid. This is not what one finds in the rocks. Lowland sloths that could not have fled into the mountains on short notice are only found in the uppermost, youngest strata. Dinosaurs and people are not found in the same rocks.

Unlike those who originally offered such ideas centuries earlier, Whitcomb and Morris made no attempt to test them against the geologic record. Instead, they questioned standard geologic evidence and, like their predecessors, invented scenarios and miracles as needed to explain inconvenient aspects of the biblical narrative. To solve the problem of getting animals to and from the ark, they argued that those making it onto the ark lived close by. After all, world geography must have been quite different before the Flood. They simply invoked supernatural assistance to cover the care and feeding of all the animals.

Whitcomb and Morris admit that the biblical flood could not have occurred before 10,000 BC, the date by when archaeological consensus then held that people had made it to North America. So they rejected carbon dating in order to conclude the archaeological dates must be wrong. In particular, they criticized the assumptions of a constant ^{14}C concentration in the atmosphere, a constant cosmic ray flux, and a constant radioactive decay rate to argue that carbon dating only worked for the time after the Flood. They explained that Earth's original vapor canopy served as a cosmic radiation shield, inhibiting the formation of ^{14}C in the atmosphere until after Noah disembarked. They then invoked greater rates of radioactive decay before the Flood to make geologic data fit a young Earth. They ignored how this would have generated tremendous heat, making paradise hellish in the days before their vapor canopy collapsed.

There is some validity to their claim that carbon dating is affected by variations in the history of Earth's atmosphere and cosmic ray activity. Cosmic ray activity does indeed vary through time—just not enough to matter all that much. Whitcomb and Morris's claim about its crippling effect on carbon dating was debunked in the 1980s, when Minze Stuiver and colleagues at the University of Washington worked out a calibration curve that extended back 13,300 years by simply counting tree rings in cross sections of logs cut at a known date and then carbon dating material from individual rings that could be lined up like overlapping bar codes from the ring patterns of different trees.

Whitcomb and Morris did not stop there, however. They argued that plants, animals, soils, and rocks were all created with the appearance of age. God made rocks with isotopic compositions identical to what one would expect had they really been ancient. In their view, the real flaw with radiometric dating was that God had put just the right amounts of different radioactive isotopes into rocks and the fossils they contained to make them seem really old.

This was not the first time that the doctrine of apparent age—the idea that God made the world to look old—was invoked to explain away geological evidence. Such thinking was popular among nineteenth-century defenders of a global flood who argued that God preloaded fossils into rocks and made them look like they had been deposited naturally. This idea that had been laughed out of Victorian England took root in cold war America.

Whitcomb and Morris even recycled Cotton Mather's arguments about antediluvian giants. Claiming that human and dinosaur footprints found along the Paluxy River near Glen Rose, Texas, were so close together that they overlapped, they included a photograph purporting to show human footprints alongside those of dinosaurs. Pointing out the tremendous size of the footprints they reminded the reader of the biblical statements about giants in the days before the Flood. However, years later, after seeing the famous tracks for himself, Morris acknowledged they were just dinosaur footprints after all.[5]

By using the Flood to explain the entire sedimentary record, Whitcomb and Morris proposed a version of geologic history that seventeenth-century cosmologists would have recognized as one of their own. Ignoring all the data that convinced eighteenth- and nineteenth-century flood supporters to give up on the idea of a global flood, Whitcomb and Morris focused on that which geologists could not explain. They thought that a great flood provided as good an explanation as geological theories if one abandoned the idea that different fossil assemblages recorded life at different times.

Whitcomb and Morris actually had some legitimate concerns and pointed out problems with the traditional views of earth scientists. What, for example, did kill off the dinosaurs? Serious objections existed to most theories of dinosaur extinction. Here was a mystery with the last chapter torn out.

Another mystery lay in the great stacks of marine sedimentary rock now stranded on continents high above the sea. How did they get there? Whitcomb and Morris noted that geologists had no explanation for this phenomenon. The only modern force with any real potential to raise up mountains was an earthquake, but the uplift observed during historical earthquakes would not add up to much change over the brief time they claimed the Bible allowed for all of earth history. As far as they were concerned, the processes that raised mountains and folded rocks were no longer operating.

Seizing on what they saw as fatal failures of conventional geology, Whitcomb and Morris revived the discredited idea of a global flood. Their case, such as it was, would soon crumble in light of plate tectonics. But geologists hadn't yet discovered the secret to the movement of continents.

Whitcomb and Morris argued that the stratigraphic order to the world's rocks that geologists had painstakingly worked out was fiction because it was based primarily on the idea of fossil succession. They thought geologists used circular reasoning in working out geologic history by interpreting the age of rocks based on the fossils they contained. This would indeed be circular reasoning—if they were right. Instead their words serve to advertise how little they bothered to learn about what they were critiquing and how they conflated geology and evolution.

In reality, the most basic aspect of the geological time scale is superposition, Steno's old idea about which rocks are above or below which other rocks. That fossil succession tracks this order has been confirmed rather than assumed. Stratigraphic relationships are strikingly clear in places like the Grand Canyon where we began our story. One does not need to look at the fossil record to understand which formation lies where in the sequence exposed in the canyon walls.

Whitcomb and Morris pointed to places where older fossil assemblages lay above younger ones as evidence that geologists just made up the stratigraphic column to fit the preconceived idea of fossil succession. But their argument ignores both regional structural mapping, which can track the deformation of folded and faulted beds across the landscape, and well-known ways to determine independently whether sedimentary beds are right side up or upside down—like how the orientation of ripple marks in sand beds or mud cracks in fine-grained rocks reveal the top and bottom of sedimentary rocks. In places where older strata lie on top of younger strata one consistently finds evidence of either folding or thrust faulting, such as upside-down beds, the fault plane itself, or a broken hash of sheared and crushed rock along the fault zone. None of these relationships depends in the slightest on the nature of the fossils that the rocks contain.

Additional ways to tell whether strata are right side up or upside down include the orientation of raindrop craters, graded bedding that records the settling out of different grain sizes (coarser material settles faster and ends up at the base of a deposit), and the orientation of burrows, which obviously extend down from what was then the surface into a deposit because overlying strata did not yet exist.

The very existence of upside-down strata presents a fatal problem for flood geology. How could sediments settle out upside down during a flood unless gravity were somehow simultaneously switching back and forth during it? If nothing much happened since the Flood, how did geological formations it laid down get flipped upside down? In contrast, given enough time, geological deformation along faults could invert rocks or shuffle the deck of rock formations as continents collided or ground past one another.

As if such concerns were not enough, fossilized coral reefs really provide the nail in the coffin for flood geology. Whitcomb and Morris explain fossil reefs found in the geologic record as ripped up and deposited along with everything else during the Flood. But if you actually go out and look at ancient reefs, as I did at my undergraduate field camp, you find that they are not composed of randomized chunks mixed up in the chaotic detritus of a violent deluge. Instead you generally find a massive limestone core, sometimes with delicate corals still in growth position. Whole reefs are preserved along with the associated lagoons, fore-reef and back-reef zones, and open-water marine environments right where you'd expect to find them in relation to one another in a modern reef. Preserving the spatial arrangement of different parts of a coral reef while ripping it to pieces and flinging them around the globe presents a logical absurdity.

Ignoring the equally awkward question of how Noah could have accommodated a coral reef on the ark, we can readily examine how long it must have taken to form modern reefs after Whitcomb and Morris's hypothesized Flood, which would have killed off living corals in a slurry of sediment-laden water. Individual corals grow at most about half an inch per year, but reefs generally grow just millimeters a year because surf incessantly pounds them. Even assuming an unreasonably generous centimeter per year growth rate, living reefs more than 1,000 meters thick would require more than 100,000 years to grow.

Additional fatal flaws have been identified in Whitcomb and Morris's ideas. Problems with their vapor canopy shrouding the early earth in a mild and uniform climate include the awkward issue that suspending even just a third of the water in the modern oceans as a vapor canopy would result in atmospheric pressure at the ground surface great enough to flatten living things like pancakes. The associated greenhouse effect would have led to runaway warming, producing a climate more like Hades than paradise.

Finally, although the Bible does not say a word about sedimentary rock or fossils, Whitcomb and Morris's own logic refutes flood geologists' central claim that sedimentary rocks did not exist before the Flood. A literal reading of the Bible requires that such rocks already existed at the time of the Flood because bitumen, the pitch or tar Noah used to caulk the ark (Genesis 6:14), comes from sedimentary rock.

Instead of grappling with these dilemmas, Whitcomb and Morris focused on challenging uniformitarianism, which they saw as the foundation for the greater evil of evolution. But they misunderstood Lyell's argument, thinking it claimed that things had always been just as they are, rather than that the underlying physical laws were constant. In Lyell's view, if you wanted to understand the types of deposits that a global flood would leave behind, you'd start by studying the deposits left by big floods. He was trying to develop a sound methodological basis for geology. Bizarrely, after ranting about how Lyell hypnotized generations of geologists, Whitcomb and Morris turned around and adopted his uniformitarian approach in arguing that hydrodynamic forces acted on the debris churned up by the Flood to sort it all out into fossil-bearing strata.

Christian reaction to *The Genesis Flood* was mixed. Some evangelical magazines praised it for its defense of Genesis, but even Whitcomb admitted that most evangelicals he knew accepted the reality of an old Earth. Yet, the book proved wildly popular among the fundamentalist rank and file, revitalizing flood geology and spawning modern creationism.

Why did Whitcomb and Morris's young-Earth creationism resonate so loudly among fundamentalists? One critic suggested that it was appealing because, unlike previous creationist books, it included footnotes and looked scholarly. Their emphasis on a plain-sense meaning of the Bible also allowed Whitcomb and Morris to present themselves as more faithful to the Bible than those who reconciled it with science through reinterpretations such as the day-age and gap theories. Their flood geology did not require them to interpret days as meaning ages or to invoke unmentioned gaps in the biblical narrative. According to Whitcomb and Morris, the Bible simply said what it meant—simply. The way they read the Bible appealed to fundamentalists.

They also gained supporters because after generations of self-imposed separatism their audience was almost entirely ignorant of modern geology. And their book appeared just as fundamentalist outrage grew heated over the widespread introduction of high school textbooks that included accounts of evolution in the post-Sputnik attempt to modernize American science education.

Whitcomb and Morris drew a direct line from geology through evolution to the communism they saw threatening Christian America. A century earlier, at the funeral of Karl Marx, Friedrich Engels invoked Darwin, crediting Marx for the discovery of the law of economic evolution. A century later, Whitcomb and Morris saw their world under threat from the rise of what they considered an amoral scientific elite that had abandoned Christianity and joined the effort to promote the socialist ideal of the common good. Geology and the evolution it supported lay at the root of the decay of modern society. Like communists, geologists, they believed, must be stopped.

Morris went on to found the Institute for Creation Research, which promotes flood geology to a lay audience through glossy publications and public lectures. With its slick propaganda machine, the Institute spearheaded the rise of young-Earth creationism and continues to influence evangelical thought.

In the mid-1960s a geologist named Davis A. Young appeared to offer Morris a ray of hope in his campaign to upend the geological establishment. The son of an eminent Old Testament scholar, Young studied geological engineering at Princeton in the late 1950s, where he flirted with accepting uniformitarianism. After enrolling in a master's program in mineralogy at Pennsylvania State University, Young read *The Genesis Flood* and became convinced that geologists needed to, once again, seek evidence in support of the Flood. Taking up the challenge, he started a PhD program at Brown University, but by 1969 he confessed to Morris that he no longer believed in a global flood. Still deeply committed to scriptural inerrancy, Young became a leading evangelical critic of young-Earth creationism.

In 1972, Morris's disappointment turned to anger when Young published a letter in a Presbyterian magazine warning that geologically illiterate creationists threatened the credibility of Christianity. Five years later, in his book *Creation and the Flood* (1977), Young went a step further and accused creationists of advocating junk science and criticized the American Scientific Affiliation for going too far in promoting biblical reinterpretation. Hoping to lead evangelical Christians to middle ground, he echoed nineteenth-century theologians in correlating earth history to the sequence of events reported in Genesis and interpreting the creation week as a figurative week in which the seventh day is ongoing.

Pointing out discrepancies in the genealogical listings presented in different books of the Bible,[6] Young maintained that the obvious interpretation of Scripture may not always be the correct one. He held that a careful reading of the Bible revealed no fundamental conflict between science and Christianity.

> *The Christian scientist is not compelled to reject the concept of the general development of the universe in accordance with physical, chemical, geological, and biological laws and processes formed by God at the very beginning and continuing to the present time.*[7]

Young accused flood proponents like Whitcomb and Morris of relying on untested speculations rooted in pure imagination and maintained that the failure of both the scientific community and mainstream theologians to engage in explaining the biblical deluge helped flood geology remain popular. Young also complained that Christians who defend traditional ideas of the Flood were too quick to appeal to miracles to help them evade scientific difficulties. It was telling how those seeking to support a global deluge consistently claimed as much scientific support as they could marshal and then invoked miracles when their own explanations broke down.

Another awkward problem for creationists lay in their claims that Noah's Flood deposited the world's sedimentary rock and that Noah landed his ark on Mount Ararat. Creationists can't have it both ways: the geologic map of Turkey shows that the stratovolcanoes forming Mount Ararat are built upon and are therefore younger than a whole series of sedimentary rocks. If the mountain itself postdates the Flood, how could Noah have landed on it? Mount Ararat itself eloquently refutes the claim that Noah's Flood was responsible for laying down all the world's sedimentary rock.

Before I read *The Genesis Flood* for myself I had been mystified as to how Whitcomb and Morris could in good faith advocate the discredited ideas that revived modern creationism. But I now see that they latched onto questions for which geologists lacked compelling answers.

In the late 1950s, geologists did not have satisfying explanations either for the relationship of continents to one another or for the origin of mountains. Nineteenth-century scientists generally thought that the breakup of the continents happened early on in earth history. Mountains were thought to have formed as the originally molten planet cooled and contracted. Continents formed in the places in which they were still found, their edges crinkling up into mountains. But the discovery that the radioactive decay of minerals produced substantial heat contradicted the theory that Earth was cooling. And no cooling meant no contracting.

Others had accepted Hutton's explanation for mountain formation. The deposition of thick sequences of sediment heated the bottom of the pile enough that its weight converted material at the bottom to rock. Somehow the heating of the sediment pile then caused uplift that formed mountains. But the discovery that oceanic crust was made of dense basalt, whereas continents were made of lighter granitic crust, meant that heating up an ocean basin couldn't turn it into a continent. Hutton's conception of the immense depth of geologic time fared better than his mountain-building theory. What then could explain the existence of mountains and the arrangement of continents?

A German meteorologist, Alfred Wegener, was the first to propose continental drift. He argued that the continents slowly moved around, sometimes colliding to join together and other times breaking apart. He thought that all the continents were originally joined in the supercontinent of Pangea (all Earth) that gradually rifted apart several hundred million years ago.

Like Bretz's flood, Wegener's unsettling idea of wandering continents was widely ridiculed when first proposed. He offered no mechanism to explain how continents split apart and then how later the pieces could come back together. His argument was based on biogeography, the global distribution of plants, animals and fossils, and the supposition that the presence of tropical fossils at high latitudes could be due to continents moving across climate zones. Similar types of fossils were found in ancient rocks on continents that have few modern species in common, which suggested the separation of once connected landmasses.

Most American geologists did not believe that Earth's crust could withstand the compressional forces required to move continents around. At a 1928 symposium held to debate the idea of continental drift, one eminent geologist accused Wegener of cherry-picking facts that supported his idea and ignoring facts and principles opposed to it. Another complained that for Wegener to be correct geologists would have to "forget everything which has been learned in the last 70 years and start all over again."[8] The conventional idea of stationary continents and ocean basins as ancient features worked well enough so that geologists did not believe they had to start from scratch in trying to explain earth history.

It took several more decades to develop a valid explanation of just how mountains arose. The answer was plate tectonics. In fact, the theory of plate tectonics came together to explain three independent mysteries that only made sense when considered together—high heat flow over submarine mountain ranges out in the middle of the oceans, a bar-code-like pattern of magnetic stripes on the seafloor, and the global distribution of earthquakes. Different groups of scientists working in different places, on different problems, with new technologies independently discovered the pieces needed to solve the grand puzzle of what caused continents to move across Earth's surface.

Before sonar, seafloor topography was veiled beneath the waves, essentially unknown in deep water away from reefs and oceanic islands. The development of sonar opened an entirely new view of the seafloor. Widely used during the Second World War as a form of underwater radar to hunt enemy submarines and help better target surface ships, sonar also could be used to map the depth of the seafloor. The idea behind sonar is simple: bounce a pulse of sound off something and measure how long it takes for the echo to come back. Knowing the speed of sound you can then determine the distance to the reflector. The process is similar to the navigation method used by bats. When oceanographers began systematically mapping the seafloor, they discovered something they could not explain. Hidden from view, under miles of water in the depths of the world's oceans, linear chains of mountains circled the world like the stitching on a baseball. If this wasn't odd enough, the spines of these ranges were always a long valley with extremely high heat flow and active rift volcanism. The ocean floor was spreading apart, generating new oceanic crust along mid-ocean ridges.

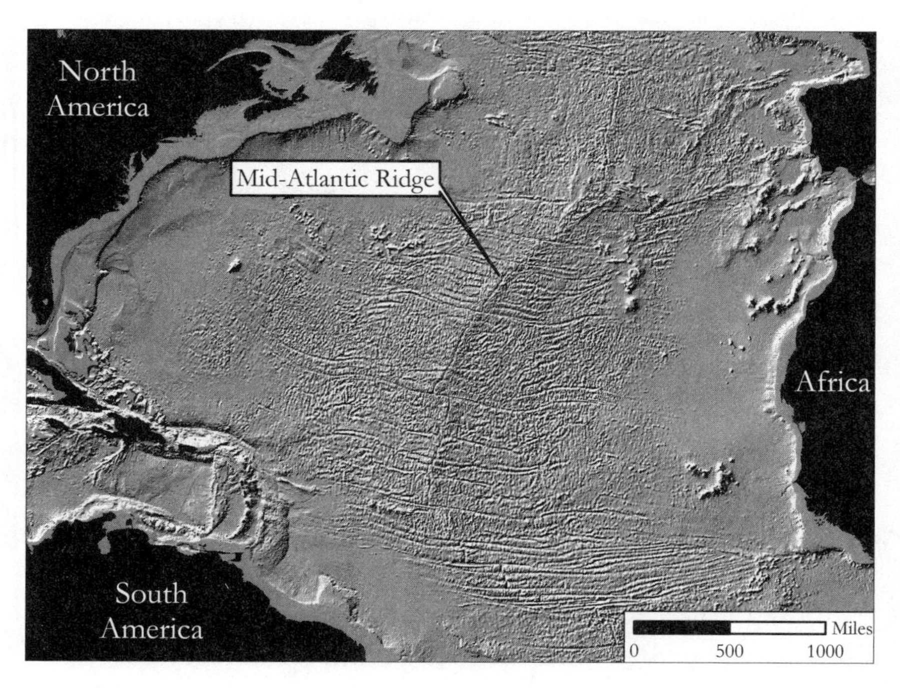

Map of the Mid-Atlantic Ridge in the North Atlantic Ocean.

The second piece of the puzzle followed the development of magnetometers that could detect subtle differences in the strength and the sign (positive or negative) of magnetic fields. Detecting the magnetic signatures of submarines proved to be an excellent way to track and sink them—if one could discern their signal from the background magnetic signature of the seafloor. For that it was useful to have a map of oceanic magnetism. The navy's program to map the magnetic signature of the seafloor produced another surprise. As researchers surveyed away from mid-ocean ridges, they found alternating bands of different magnetic polarization—stripes with normal magnetic polarity alternating with stripes of negative polarity. The scientists who first reported the odd zebra-hide pattern of magnetically striped seabed sections noted that they had no idea how to explain it. A decade later, the stripes were understood to be like a strip-chart recording of global reversals in Earth's magnetic field. Imprinted with the planet's magnetic field—positive or negative—when they first cooled, the newly formed rocks at the mid-ocean ridges slowly moved away from the ridge as new crust squeezed up at the ridge and pushed older crust out of the way.

The discovery of how and where earthquakes are produced provided the final critical piece of the tectonic puzzle through efforts to verify nuclear test ban treaties. The discovery that underground nuclear tests produced seismic waves that could be distinguished from the seismic waves produced by earthquakes fueled substantial government investment in seismology in the 1950s and 1960s. Ratification of the Limited Test Ban Treaty in 1963 made continuous seismic monitoring and locating of earthquakes critical to verifying treaty compliance. Establishment of a global seismograph network led to a rapid increase in the global catalog of earthquake locations, revealing a striking pattern. Most earthquakes occurred in the uppermost several hundred miles of Earth's crust. But there was no good explanation for mysterious deep earthquakes triggered more than four hundred miles below ground. The rocks that far down should be too hot and mushy to sustain the rigid deformation needed to produce earthquakes. As seismologists refined their methods, they found that these unusually deep earthquakes outlined slabs of crust sinking down into the planet's interior.

These three seemingly unrelated observations—the birth of new oceanic crust at spreading centers in the middle of the oceans, the slow movement of new crust away from the spreading centers, and the sinking of crust beneath continents in subduction zones along the ocean margins—defined a full cycle. New crust was rising up in the middle of the ocean and sinking back down to be recycled in the deep trenches at the edge of ocean basins. Crust produced at mid-ocean ridges was being pushed aside until it ran into a lighter continent and got shoved back down into the mantle. Continents rafted along on mobile plates of crust spread apart and collided with each other, pushing up mountains and continually resurfacing the planet over geologic time. Here was a single, grand mechanism to explain not only how continents moved around but how mountains formed, where different rock types occurred, and why earthquakes, volcanoes, and mountains all line up where plates split apart, collide, or slide past one another.

In contrast to the long list of fundamentally irreconcilable problems with the concept of a global flood—such as where the water came from and where it went after the flood, the order of the fossil record, the predominance of extinct species in the fossil record, and the presence of soils and burrows developed on sedimentary rocks—plate tectonics provides remarkably consistent explanations for a wide range of phenomena. It explains why Africa and South America look like they fit together like the pieces of a jigsaw puzzle, only separated by the Atlantic Ocean. It explains the sequences, ages, and assemblages of rocks one finds throughout the world, as well as the global distribution of topography. Plate tectonics revolutionized

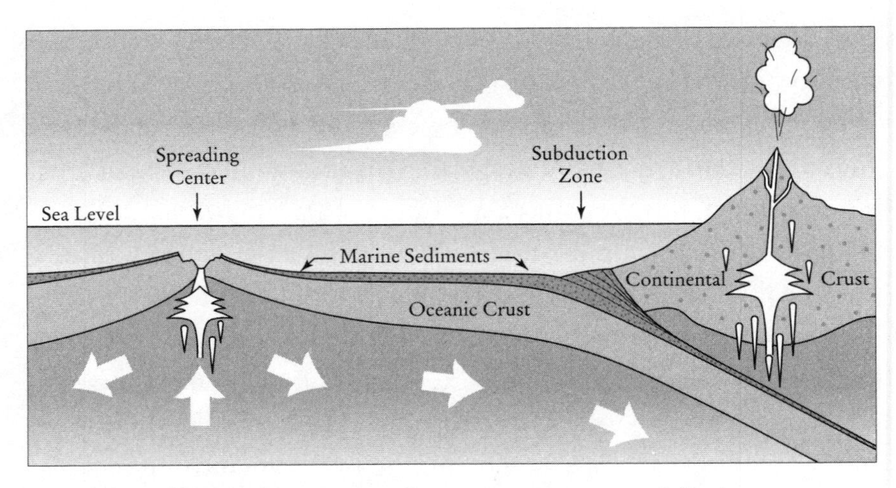

Illustration of how mid-ocean spreading centers connect to subduction zones to define the upper half of deep Earth circulation that drives plate tectonics and rafts continents along as oceanic crust moves laterally (based on a sketch by Véronique Robigou).

geology, elegantly explaining within a single framework many of the physical world's outstanding mysteries.

In the early 1970s, as plate tectonics was revolutionizing the earth sciences, James Moore, a lecturer in history of science and technology at England's Open University, emphasized the ongoing threat dogmatic theology posed to evangelical thought in the *Evangelical Quarterly*.

> *Scientists and theologians who are Christians too often neglect history to their own intellectual condemnation. What appear to them to be major modern issues on which turns the perspicuous truth of biblical revelation are often problems which were long ago laid to rest. They realize too late that their labored polemics and hastily written tracts are little more than exuberant rehashes of what was once conclusively argued or contravened. This is unforgivable. They should have known that, without historical sense, it is not only possible but inevitable that men repeat the very errors which once discredited their forbearers.*[9]

Like their seventeenth-century predecessors, the imagination of modern creationists knows no bounds. The impressive variety of explanations they invoke for substantiating literal interpretations of Noah's Flood as a global catastrophe include collapse of a globe-shrouding vapor canopy, eruption of supergeysers from the earth's core, and bombardment by asteroids striking the world's oceans. Apparently, a literal reading of the Bible still leaves a lot of room for creative interpretation.

Rocks, on the other hand, do not. When a colleague and I were leaving the Creation Museum, he pointed out the 440-million-year-old, trilobite-bearing limestone exposed in a road cut. This ancient seabed from the proto-Atlantic Ocean is exposed for hundreds of miles across Ohio, Kentucky, and Indiana. These rocks are clearly not a chaotic, mixed-up product of an earth-shattering flood. They formed when an ancient sea gradually accumulated, layer by layer, to form a thick pile of sediment stretching from Newfoundland to Alabama. What better testimony could be offered to the irreconcilable differences between geology and creationism than the fact that the Creation Museum itself is built upon rocks that dispute the version of earth history displayed within its walls?

But without a theoretical framework for interpreting their observations, geologists can misread the stories inscribed in the rock record. Only when the idea of plate tectonics came to light was there a clear driving mechanism to explain the global distribution of topography and rock types. The difference between scientists and creationists, however, is that scientists assess their theories based on how well they fit the evidence, whereas creationists interpret observations by determining how well the facts fit in with their beliefs. Not surprisingly, these different varieties of faith yield radically different views of nature.

13

The Nature of Faith

THE PUSH-AND-PULL, the back-and-forth through history between science and religion is more of a dance than a war. I now think of it as an awkward egalitarian waltz, with the partners trading off the lead, sometimes moving one step ahead, other times following behind, and occasionally stomping on each other's toes. Science and religion share humanity's strong desire to understand our world, and as is the case with Noah's Flood, much of the conflict we perceive between them occurs over how to interpret ancient stories in light of modern knowledge.

What more can we learn from the story of Noah's Flood? Even though we can no longer read the story literally, we can still learn from it—all of us. The story of the biblical flood story shows how it is as important for scientists to maintain flexibility of interpretation when facing new data as it is for theologians who don't want to be backed into making implausible arguments, like having to claim that rocks lie.

In our journey through the history of flood stories, we have seen two different ways of viewing faith—faith in a method or process (like science) and faith in a particular idea, view, or conclusion (like scientific theories or religious beliefs). Based on the idea that open inquiry promotes learning, scientists put evidence first to formulate and build upon theories. Evidence that does not fit a theory is scientific gold—it leads to new discovery and knowledge. In contrast, elevating preconceived ideas and beliefs above evidence shuts off learning and stifles curiosity about the world. Here lies the question at the border between science and religion. Is seeing believing, or is believing seeing?

Perhaps the conventional view of the centuries-long conflict between geology and Christianity should be viewed as more of a coevolutionary process. As geological interpretations of earth history evolved away from and evidence mounted against a single, worldwide deluge, Christians responded in three ways: some abandoned the Bible as an authoritative text; some tried to reconcile biblical and scientific views; and others simply rejected evidence they perceived to threaten biblical authority. These approaches correspond, respectively, to secular modernism, mainstream Christianity, and reactionary fundamentalism.

To a geologist like myself, interpreting ancient stories of great floods presents an alluring challenge. I would hope that we can all appreciate how, after centuries of debate and creative explanations, it appears that humanity's rich legacy of flood stories reflects a variety of ancient disasters. The global pattern of tsunamis, glacial outburst floods, and catastrophic flooding of lowlands like Mesopotamia or the Black Sea basin fits rather well the global distribution and details of flood stories. Considered together, geological and anthropological evidence suggests rational explanations for why flood stories are uncommon in Africa, why they are so different in China, and why they are widespread in the Middle East, northern Europe, America, and all across the Pacific. Time and again, great floods swept worlds away in disasters remarkable enough to shape humanity's oldest stories, which were then passed down through generations—and civilizations—to become powerful legends.

Every day at work, I walk past a slab of polished rock that elegantly refutes the idea of Noah's Flood as being the sole event of earth history. It hangs on the wall down the hall from my department's office and is a gorgeous tableau of colorful sedimentary rocks embedded within different sedimentary rock—a stone tablet made from cobbles, gravel, and sand. Like Steno's solid within a solid, this conglomerate shows that at least two grand catastrophes or geological cycles played a role in earth history. One cannot stand before it and embrace young-Earth creationism and its single world-wrecking flood without abandoning faith in earth history told by the rocks themselves.

Contrary to creationist claims, reading the geologic record does not depend on paleontology and evolution—they provide complementary constraints on earth history. The astounding degree of agreement between the geologic and fossil records would require miracles upon miracles, were it not simply indicative of the fact that they independently recorded the same grand story.

We need a historically informed understanding of how people read and interpreted sacred texts in the past in order to inform how we read them today. It is as wrong-headed for atheists to assume that religion demands that the faithful read biblical stories literally as it is for the faithful to use scripture to bash modern science. At the same time, those who seek to reconcile science and religion need to confront the intellectual problem of miracles. These different ways of investigating truth come into direct conflict when scientific findings contradict religious beliefs. After all, the creationist view of the week of Creation and Noah's Flood as a comprehensive record of earth history leaves no room for central discoveries of modern geology like plate tectonics and the realization that long periods of time are required for erosion to sculpt the land.

Throughout church history, biblical commentators from across the theological spectrum used extrabiblical information about the natural world to help interpret the story of Noah's Flood. Naturally, most wanted to explain their world in terms they understood. After all, who needs a confusing universe? As knowledge of the world grew, so too did explanations for what shaped it. Along the way, concepts of how landscapes evolved changed in ways that paralleled developments in Christian theology. Early conviction that the world was slowly decaying gave way to creative schemes like those involving violent catastrophes to generate Noah's Flood, and then to grand cycles of repeated catastrophes that destroyed multiple worlds. Finally, the modern concept emerged—a planet on which life, land, and the atmosphere are intimately interconnected and self-renewing through plate tectonics, a process that continually remakes the world over unimaginably deep time.

Before the Reformation, theologians generally agreed that simple folk should accept the Bible as literal fact but that learned persons who could read the original Hebrew and Greek texts might discover deeper meanings. The Protestant idea that anyone could read and interpret the plain words of the Bible for themselves led to a flowering of divergent interpretations. Theological arguments about Noah's Flood evolved as scientific theories reframed rational explanations and Christians reinterpreted Genesis to accept the knowledge that Earth had a long and dynamic history. It was seen as fruitless to debate the very rocks that made up our world.

That modern creationism is one of the most recently evolved forms of Christianity may surprise today's fundamentalists. Yet before the rebirth of young-Earth creationism in the 1960s, most fundamentalists subscribed to either the gap or day-age theories that fit geologic time into the opening verses of Genesis.

In fact, the founding fundamentalists did not want to choose between science and religion. One, they believed, leads to greater understanding and knowledge about the way the world works, and the other provides moral and spiritual guidance in navigating the complexities of life, culture, and society. Seen in this light, the varying interpretations of the biblical flood story are part of an ongoing battle for the soul of Christianity. Will it remain a dynamic faith that helps people navigate modern times and understand the world and our place in it? Or will Christianity become locked in a senseless war against reason, as St. Augustine feared? Only time will tell.

Together with mankind's relationship to the environment, the relationship between science and religion is one of the most important—and difficult—problems facing humanity today. Not surprisingly, these problems are linked. Herein I see another modern lesson of the Noah's Flood story. Perhaps we would be wise to consider Earth itself as a habitable ark careening around the Sun. Maybe the modern relevance of the story lies not so much in whether it literally describes a particular prehistoric flood, but in a timeless lesson about humanity's moral responsibility to safeguard creation, as did Noah and his crew.

To me, a literal reading of the Creation in Genesis does not do the story justice. Even a casual reading reveals that days one, two, and three set the stage for days four, five, and six. The creation of light on the first day sets the backdrop for the creation of the sun, moon and stars on the fourth day. The separation of the sky from the waters on the second day sets up the creation of birds and creatures of the sea on the fifth day. The segregation of dry land on the third day sets the stage for the creation of plants and terrestrial animals on the sixth day. This recurring cycle of three is a classic poetic device. I don't think that the Creation story was intended as historical fact. It's more akin to epic poetry written to convey the divine origin of our wondrous world and everything in it, however they came about. Genesis 1 remains powerful and relevant today if read as a symbolic polemic intended for early monotheists rather than as a Bronze Age scientific treatise.

One challenge of interpreting Genesis literally lies in its brevity. The Creation is described in the fifty-six verses of Genesis 1–2. Noah's Flood is covered in the sixty-eight verses of Genesis 6–8. In other words, about all there is in the Bible to explain the 4.5 billion years of earth history is about the same number of sentences on a typical front page of the *New York Times*. One can hardly expect a detailed accounting given that this represents just a couple of dozen sentences per billion years, and that most of these sentences deal with the life and times of Adam, Noah, and company.

One might think that brevity equates with clarity in a simple literal reading of Genesis. But God created light on the first day and didn't make the Sun until the fourth day. So where did the light and the night come from, and how was the length of the first three days defined? Fish were not even created at all in a literal reading of Genesis, for they are not mentioned. Neither are bacteria, viruses, and insects—or dinosaurs. Does this mean that they evolved after the initial Creation, or that Genesis is not a comprehensive world history? Such questions and the potential for alternative interpretations gave rise to a long history of commentary on how to interpret Genesis, and how to interpret the story of Noah's Flood in particular.

Perhaps the challenge of interpreting another famous document—the United States Constitution—can help illuminate the problem of trying to understand the Bible. Consider how little liberals and conservatives agree on the meaning of the Constitution, a document only a few thousand words long, written in English not that long ago, whose signed original is on display under glass for all to see. Compare that with the Bible, which was pieced together from partial versions of a work three-quarters of a million words long, handed down between cultures, and translated several times over from a language lacking vowels and spaces between words. Is it any surprise that people today don't agree on exactly what the Bible means?

Like most geologists, I had come to see Noah's Flood as a fairy tale—an ancient attempt to explain the mystery of how marine fossils ended up in rocks high in the mountains. Now I've come to see the story of Noah's Flood like so many other flood stories—as rooted in truth. But was it the flooding of the Black Sea, or a great Mesopotamian flood that ravaged the ancestral homeland of Semitic peoples? Who knows? I doubt the historic truth about Noah's Flood will ever be known with certainty. And I don't think it really matters. The discoveries of science have revealed the world and our universe to be far more spectacular than could have been imagined by Mesopotamian minds. To still see the world through their eyes is to minimize the wonder of creation.

Our interpretation of the world around us fundamentally shapes our outlook. We will only look for evidence that confirms our beliefs if we have already decided how and what to think about something. But if we keep our minds open, we may be surprised at what we discover. And how we choose to view the world seems to increasingly frame contemporary issues of tremendous societal importance, from climate change to the way we teach science in public schools. At stake is how we interpret nature, and what, if anything, we can learn from the world around us.

Geologists make sense of ancient events by piecing together stories archived in stone and inscribed on the land; we attempt to forge coherent theories that stand up to evidence. Most attempts fail. But that's central to an ongoing process of pushing old theories until they break in order to improve upon them. Yet, we've seen how the scientific establishment can be inherently resistant to change, favoring familiar theories over new or uncomfortable ideas. What distinguishes science from religion is that in science even cherished ideas must stand up to the test of new evidence.

By design, science excludes miracles because there is no way to test them through rational analysis. Science cannot address supernatural or divine action any more than Seattle residents can will away gray skies. Creationists and advocates of intelligent design seize upon this fundamental limitation of the scientific method to allege that science denies the existence of God. But science can no more prove God does not exist than it can prove He (or She) does exist. And no matter how much we learn about the material characteristics, properties, and history of the universe, such knowledge will not explain why the universe exists or how it came to have the properties it does. This will always be a matter of speculation—or faith.

However, we cannot simply compartmentalize science and religion into tidy, noncompeting domains because some scientific discoveries are not compatible with particular religious beliefs. Few religious ideas can be tested, but some are refutable. Science has demonstrated that once-conventional beliefs concerning the physical world are wrong—like the ideas that we live at the center of the universe on a six-thousand-year-old planet shaped by Noah's Flood. I believe faith and science can peacefully coexist, so long as we don't founder on or cling to the rocky shore of either. What this requires is open-minded thinking guided by humanity's greatest asset—the gift of reason.

Naturally, there is bound to be some friction between science and religion because they offer very different ways to assess truth. The long history of interaction between geology and Christianity includes times when they reinforced one another and times when they clashed. The story of Noah's Flood shows how the different beliefs of various branches of Christianity are shaped by which parts of the Bible their devotees read literally and which they interpret allegorically. Over time, Christian thought has sorted itself out along a continuum of belief. The modern view of inherent conflict is championed most vociferously by those who keep the conflict going—creationists and militant atheists who share little else than the belief that faith in God and science are incompatible. Most people, however, hold beliefs somewhere between these two extremes.

In reality, there is a wide spectrum of possible beliefs about the relationship of God to the material world. At one end is belief in an engaged, helpful personal God who rides shotgun on everyday activities and can intervene at anytime to favor the outcome of specific events, like a coin toss or a football game. Others believe in a more strategic God that intervenes only occasionally to shape the course of history or important events, like elections or wars. Farther along the continuum of belief is a more distant God responsible for creating the universe and the laws governing the world. At this end of the philosophical spectrum are the beliefs that God directed and planned the course of events in advance, and the view that the universe is a glorious but random experiment. Still others ascribe no role in the universe for a God at all.

While religion cannot adequately address scientific questions, accepting scientific truths need not mean abandoning morality, purpose, and meaning in life. And just because science can neither prove nor disprove the existence of God does not mean that it says religious faith is an illusion. Thoughtful discussions of the relationship between science and religion are impossible when fundamentalists disguise religious arguments as science and scientists dismiss religion as childish superstition. In reality, faith and reason need not be enemies if one views ignorance as the enemy of both. Should humans be afraid of an enigmatic universe whose mysteries elude us? Or should we struggle to decipher the mysteries of our world and how it works, whether for simple intellectual joy and challenge, to reap practical benefits, or to gain insight into the mind of God—whatever one imagines that to be.

Geologists have uncovered a grand story of the coming and going of life and the making and remaking of whole worlds as continents wandered the globe over billions of years. We are still unraveling the secrets behind the great extinction events of the past and learning to understand the causes of ancient climate changes that ushered in times hotter than today and periods when the whole planet froze over. Even now, as we send robots off to explore the geology of Mars, our nearest celestial neighbor, we are discovering new planets circling distant stars. We will likely soon confront the discovery of other inhabitable planets in a universe far grander than ever imagined in our intellectual infancy.

The scientific story of the origin and evolution of life, the vast sweep of geologic time, and the complexity of the processes that shaped the world we know today inspire more awe and wonder than the series of one-off miracles from Genesis that I read about in Sunday school. Miracles do not fuel curiosity or innovation. If we embrace the claim that Earth is a few thousand years old, we must also throw out the most basic findings of geology, physics, chemistry, and biology. The concept of geologic time, on the other hand, opens up an entirely new creation story, along with the idea that the world is unfinished and creation is ongoing. And a complex, evolving world is one we would be well advised to do our best to understand. Personally, I find a world that invites exploration and learning more inspiring than a world where all is known.

While science has much to offer us, from vaccines to space travel, religion can help humanity frame essential social, moral, and ethical decisions, such as those arising from the development and uses of science and technology. Of course, history is also replete with examples of religion being used to subjugate, control, and persecute. Ethics and morality do not require a religious basis any more than vociferous professions of religious belief guarantee ethical or moral behavior. Faith and reason offer different lenses through which people seek to understand the world and our place in it.

I find that the wonder in reading rocks and topography, and in understanding the vast scope of geologic time, rivals that of religious belief. In either one can find a taste of the infinite and of things far grander than ourselves. Yet no honest search for truth can deny geological discoveries—not when Earth's marvelous story is laid out for all to see in the very fabric of our world. We may argue endlessly about how to interpret the Bible, but the rocks don't lie. They tell it like it was.

Notes

1. Buddha's Dam

1. Atoms of the same element have the same number of protons, but different isotopes of an element have different numbers of neutrons. For example, atoms of carbon-12 have six protons and six neutrons, whereas atoms of carbon-14 have six protons and eight neutrons. Different isotopes of an element therefore have different atomic mass, which allows their relative abundance to be measured in a mass spectrometer.

2. Polls reporting the widespread acceptance of creationist ideas among the American public include: a 2001 National Science Foundation survey of science literacy that found more than half of American adults did not know that dinosaurs went extinct before people walked the earth; a 2004 ABC News Poll that reported more than half of Americans believed that the biblical account of the creation was "literally true," and that Noah's Flood was a global flood; and a 2005 Gallup Poll (August 5-7, 2005) that reported more than half of Americans believed that "creationism" was definitely or probably true.

2. A Grand Canyon

1. When the presidents of the Geological Society of America and the American Geophysical Union urged the Park Service to stop selling the book, political appointees in Washington overruled the park superintendent's decision to pull it from the shelves. Instead, the book moved to its own specially created inspirational reading section. In defending the continuing sale of the book, Park Service spokesperson Elaine Sevy was quoted by National Center for Science Education deputy director Glenn Branch as saying, "Now that the book has become quite popular, we don't want to remove it" (Branch, 2004).

2. Igneous rocks form by cooling from hot magma, whether below ground (intrusive rocks) or when erupted out of a volcano (extrusive rocks). Metamorphic rocks form when preexisting rocks get heated enough under enough pressure that their minerals are transformed (metamorphosed) into new minerals and deform, sometimes to the point where they flow like taffy and produce wild swirling patterns cast in stone.

3. Bones in the Mountains

1. There is some controversy about the height and name of the mountain. While new technology has allowed more accurate measurements, the mountain also has been rising since it was first surveyed. In May 1999, the former official height of Mt. Everest of 29,029 feet was revised upward to 29,035 feet based on a multireceiver global positioning system (GPS) survey. The name of the mountain is a bit more complicated because several cultures have a claim on it. In 1865, the Royal Geographical Society named the peak after the British surveyor general of India, Sir George Everest, who first recorded a surveyed height and location of the mountain. He called it peak XV, peak fifteen, because at the time local names were not known to the British due to the area's being off-limits to foreigners. The far older Tibetan name for the mountain is Chomolungma, which I've seen variously translated as Saint Mother, Holy Mother, Goddess Mother of Mountains, or Mother Goddess of the Earth. Its Nepali name is Sagarmatha, literally sky head or Goddess of the Sky. The most recent, Chinese name is Mount Qomolangma, a transliteration of the traditional Tibetan name.

2. Origen, 1966, 288.
3. Augustine, 1982, 47-48.
4. White, 1910, 8.
5. Luther, 1960, vol. 2, 65.
6. Ibid., vol. 2, 93.
7. White, 1910, 126.
8. Drake, 1957, 181.
9. Ibid., 186.
10. White, 1910, 137.

4. World in Ruins

1. Cutler, 2003, 59.
2. Burnet, 1684, 140.
3. Ibid., 18.
4. Ibid., a2.
5. Nicholson, 1997, 235.
6. Davies, 1969, 73.
7. Burnet, 1684, a.
8. Woodward, 1723, 105.
9. Ibid., 105-6.
10. Cutler, 2003, 178.
11. Cohn, 1996, 135.
12. Keill, 1698, 26.
13. Ibid., 58.
14. Ibid., 151.

5. A Mammoth Problem

1. Cohn, 1996, 88.
2. Levin, 1988, 762, 764.

3. Mammoths roamed both northern Asia and North America, while the somewhat smaller mastodons that grew up to just ten feet in height only lived in North America. Fossils of the two species can be distinguished based on their

distinctively different teeth. Grassland-dwelling mammoths had molars characterized by long ridges that could cut through grass like scissors, whereas the woodland-dwelling mastodons had multipointed molars designed to crush leaves, twigs, and bark.

4. Cuvier, 1978, 15.
5. Ibid., 16.

6. The Test of Time

1. White, 1910, 215.
2. Playfair, 1805, 73.
3. Hutton, 1788, 304.
4. Kirwan, 1799, 105.
5. Playfair, 1802, 351, 401.
6. Ibid., 471, 472-473.

7. Catastrophic Revelations

1. Klaver, 1997, 19.
2. Cuvier, 1978, 171.
3. Buckland, 1820, 23-24.
4. Ibid., 20.
5. Ibid., 146.
6. Sedgwick, 1825, 35.
7. Buckland, 1837, vol. 1, 22.
8. Ibid., 18.
9. Ibid., 35.
10. Klaver, 1997, 19.
11. Ibid., 25.
12. Lyell, 1833, 6.
13. Ibid., 270.
14. Wilson, 1972, 310.
15. Klaver, 1997, 49.
16. Ibid., 113.
17. Sedgwick, 1834, 313.

8. Fragmented Stories

1. Smith, 1876, 4.
2. Genesis 1:21; translated as "whales" in the King James Bible.

9. Recycled Tales

1. Allen, 1963, 43.
2. Paine, 1824, 90.
3. The King James Bible mentions unicorns nine times based on the translation of the Hebrew word re'em, which was translated as monoceros (one horn) in the Greek Bible and then as unicornis in the Latin Bible (Numbers 23:22, 24:8; Deuteronomy 33:17; Job 39:9, 39:10; Psalms 22:21, 29:6, 92:10; Isaiah 34:7). While re'em is now translated as "wild ox" in most other English versions of the Bible, it remains unclear whether the original description of an untamable animal with great strength was meant to refer to a rhinoceros or an auroch, the now extinct wild ancestor of the modern cow that in antiquity was generally depicted in profile as having a single horn.
4. Zimmern, 1901, 60.
5. Ryle, 1892, ix.
6. Ibid., 112-13.
7. Lenormant, 1883, 488.
8. Frazier, 1918, 335, 359.
9. Lewis, 2006, 30.
10. Ibid., 31.
11. Huxley, 1893, 215.

10. Dinosaurs in Paradise

1. Numbers, 1982, 74.
2. Numbers, 1982, 540.
3. Schofield, 1917, 3.
4. Miller, 1922, 701, 702, 703.
5. Ibid., 1922, 702.
6. Price, 1923, 280.
7. Numbers, 1982, 540.

8. Ramm, 1956, iii.
9. Ibid., 32.
10. Ibid., 177.

11. The Heretic's Flood

1. Bretz, 1978, 2.
2. Baker, 1978, 14.
3. Bretz, 1978, 1.

12. Phantom Deluge

1. Whitcomb and Morris, 1961, preface to the sixth printing.
2. Ibid., 118.
3. Ibid., 214.
4. Ibid., 124.
5. See Numbers, 1992, 265-67, and Mayor, 2005, 339-41, for further discussion of this and other fossil frauds perpetrated by creationists.
6. Like that between Matthew 1:17, which states that there were fourteen generations from Abraham to David, and 1 Chronicles 1-2, where thirteen generations are listed.
7. Young, 1977, 106.
8. Chamberlain, 1928, 87.
9. Moore, 1973, 141.

Sources

Aksu, A. E., et al. "Persistent Holocene Outflow from the Black Sea to the Eastern Mediterranean Contradicts Noah's Flood Hypothesis." *GSA Today* 12, no. 5 (2002): 4–10.

Allen, D. C. *The Legend of Noah: Renaissance Rationalism in Art, Science, and Letters.* Urbana: University of Illinois Press, 1963.

Allen, J. E., M. Burns, and S. C. Sargent. *Cataclysms on the Columbia.* Portland, OR: Timber Press, 1986.

Anon. "The scientific meeting at York." *Chambers' Edinburgh Journal.* New Series, vol. 1, no. 47 (Nov. 23, 1844): 322–23.

Attridge, H. W., ed. *The Religion and Science Debate: Why Does It Continue?* New Haven, CT: Yale University Press, 2009.

Augustine. *The Literal Meaning of Genesis.* Trans. J. H. Taylor. New York: Newman Press, 1982.

Arnold, J. R., and W. F. Libby. "Age Determinations by Radiocarbon Content: Checks with Samples of Known Age." *Science* 110 (1949): 678–80.

Baker, V. R. "The Channeled Scabland: A Retrospective." *Annual Review of Earth and Planetary Sciences* 37 (2009): 393–411.

——. "The Spokane Flood Controversy." In *The Channeled Scabland: A Guide to the Geomorphology of the Columbia Basin, Washington*, edited by V. R. Baker and D. Nummedal, 3–15. Washington, DC: National Aeronautics and Space Administration.

Barber, E. W., and P. T. Barber. *When They Severed Earth from Sky: How the Human Mind Shapes Myth.* Princeton, NJ: Princeton University Press, 2004.

Berger, W. H. "On the Discovery of the Ice Age: Science and Myth." In *Myth and Geology*, edited by L. Piccardi and W. B. Masse, 271–78. Special Publication 273. London: Geological Society, 2007.

Branch, G. "Flood Geology in the Grand Canyon." *Reports of the National Center for Science Education* 24, no. 1 (January–February 2004).

Bretz, J H. "The Spokane Flood Beyond the Channeled Scabland." *Journal of Geology* 33 (1925): 97–115, 236–59.

——. "Bars of Channeled Scabland." *Geological Society of America Bulletin* 39 (1928): 643–701.

——. Introduction to *The Channeled Scabland: A Guide to the Geomorphology of the Columbia Basin, Washington*, edited by V. R. Baker and D. Nummedal, 1–2. Washington, DC: National Aeronautics and Space Administration.

Brown, E. H., R. S. Babcock, M. D. Clark, and D. E. Livingston. "Geology of the Older Precambrian Rocks of the Grand Canyon: Part I. Petrology and Structure of the Vishnu Complex." *Precambrian Research* 8 (1979): 219–41.

Buckland, W. *Vindiciae Geologicae; or the Connexion of Geology with Religion Explained, in an Inaugural Lecture Delivered Before the University of Oxford, May 15, 1819, on the Endowment of a Readership in Geology by His Royal Highness the Prince Regent.* At the University Press for the author, Oxford, 1820.

——. *Reliquiae Diluvianae; or, Observations on the Organic Remains Contained in Caves, Fissures, and Diluvial Gravel, and on Other Geological Phenomena, Attesting the Action of an Universal Deluge.* London: John Murray, 1823.

——. *Geology and Mineralogy, Considered with Reference to Natural Theology.* Philadelphia: Carey, Lea and Blanchard, 1837.

Burnet, T. *The Theory of the Earth: Containing an Account of the Original of the*

Earth and of All the General Changes Which it Hath Already Undergone, or is to Undergo, Till the Consumation of All Things, The Two First Books Concerning the Deluge, and Concerning Paradise. Printed by R. Norton, for Walter Kettilby, at the Bishops-Head in St. Paul's Church-Yard, 1684.

Calvin, J. *Commentaries on the First Book of Moses Called Genesis.* Trans. Rev. J. King, v. 1. Grand Rapids, MI: Wm. B. Eerdmans Publishing Company, 1948.

——. *Institutes of the Christian Religion.* Ed. J. T. McNeill, trans. F. L. Battles. Philadelphia: The Westminster Press, 1960.

Chamberlin, R. T. "Some of the Objections to Wegener's Theory." In *The Theory of Continental Drift: A Symposium on the Origin and Movement of Land Masses both Inter-Continental and Intra-Continental, as Proposed by Alfred Wegener,* 83–87. Tulsa, OK: The American Association of Petroleum Geologists, 1928.

Chorley, R. J., A. J. Dunne, and R. P. Beckinsale. *The History of the Study of Landforms or the Development of Geomorphology, Volume One: Geomorphology Before Davis.* London: Frome and Methuen & Co Ltd and John Wiley & Sons, Inc., 1964.

Clark, E. *Indian Legends of the Pacific Northwest.* Berkeley: University of California Press, 1953.

Clark, R. E. D. "The Black Sea and Noah's Flood." *Faith and Thought: Journal of the Transactions of the Victoria Institute* 100 (1972): 174–79.

Cohn, N. *Noah's Flood: The Genesis Story in Western Thought.* New Haven, CT: Yale University Press, 1996.

Colenso, J. W. *The Pentateuch and the Book of Joshua Critically Examined, Part IV.* London: Longman, Green, Longman, Roberts, & Green, 1864.

Cressman, L. S. et al. "Cultural Sequences at The Dalles, Oregon," *Transactions of the American Philosophical Society* 50 (1960): 1–108.

Cutler, A. *The Seashell on the Mountaintop.* New York: Dutton, 2003.

Cuvier, G. *Essay on the Theory of the Earth.* Ed. C. C. Albritton Jr., trans. R. Kerr. New York: Arno Press, 1978.

Dalley, S. *Myths from Mesopotamia: Creation, The Flood, Gilgamesh, and Others.* Rev. ed. Oxford: Oxford University Press, 2000.

Davies, G. L. *The Earth in Decay: A History of British Geomorphology 1578 to 1878*. New York: American Elsevier Publishing Company, Inc., 1969.

Degens, E. T., and D. A. Ross. "Chronology of the Black Sea Over the Last 25,000 Years." *Chemical Geology* 10 (1972): 1–16.

Desmond, A. J. "The Discovery of Marine Transgression and the Explanation of Fossils in Antiquity." *American Journal of Science* 275 (1975): 692–707.

Drake, S., trans. Letter to the Grand Duchess Christina (1615). In *Discoveries and Opinions of Galileo*, 175–216. Doubleday & Company, New York, 1957.

Dundes, A., ed. *The Flood Myth*. Berkeley: University of California Press, 1988.

Echo-Hawk, R. C. "Ancient History in the New World: Integrating Oral Traditions and the Archaeological Record in Deep Time." *American Antiquity* 65 (2000): 267–90.

Eells, M. "Traditions of the 'Deluge' among the Tribes of the North-West." *The American Antiquarian* 1 (1878): 70–72.

Fleming, J. "The Geological Deluge, as Interpreted by Baron Cuvier and Professor Buckland, Inconsistent with the Testimony of Moses and the Phenomena of Nature." *Edinburgh Philosophical Journal* 14 (1826): 205–39.

Frazer, J. G. *Folk-Lore in the Old Testament*. London: Macmillan and Co., 1918.

García Martínez, F., and G. P. Luttikhuizen, eds. *Interpretations of the Flood*. Leiden, Boston, and Köln: Brill, 1998.

Gaster, T. H. *Myth, Legend, and Custom in the Old Testament*. New York: Harper & Row, 1969.

Gilbert, M. T. P., et al. "DNA from Pre-Clovis Human Coprolites in Oregon, North America." *Science* 320 (2008): 786–89.

Giosan, L., F. Filip, and S. Constatinescu. "Was the Black Sea Catastrophically Flooded in the Early Holocene?" *Quaternary Science Reviews* 28 (2009): 1–6.

Goodman, D. C., ed. *Science and Religious Belief, 1600–1900*. Dorchester, UK: The Open University Press, 1973.

Gupta, S., J. S. Collier, A. Palmer-Felgate, and G. Potter. "Catastrophic Flooding Origin of Shelf Valley Systems in the English Channel." *Nature* 448 (2007): 342–45.

Halley, E. "Some Considerations about the Cause of the Universal Deluge, Laid before the Royal Society, on the 12th of December 1694." *Philosophical Transactions of the Royal Society* 33 (1724): 118–23.

Hodges, K. V., et al. "Simultaneous Miocene Extension and Shortening in the Himalayan Orogen." *Science* 258 (1992): 1466–70.

Huggett, R. *Cataclysms and Earth History: The Development of Diluvialism.* Oxford: Clarendon Press, 1989.

Hutton, J. "Theory of the Earth." *Royal Society of Edinburgh Transactions* 1 (1788): 209–304.

Huxley, T. H. *Science and Christian Tradition*, London: Macmillan, 1893.

Ilg, B. R., K. E. Karlstrom, D. P. Hawkins, and M. L. Williams. "Tectonic Evolution of Paleoproterozoic Rocks in the Grand Canyon: Insights into Middle-crustal processes." *Geological Society of America Bulletin* 108 (1996): 1149–66.

Jacobs, M. *Kalapuya Texts.* University of Washington Publications in Anthropology, vol. 11. Seattle: University of Washington, 1945.

Jahn, M. E. "Some Notes on Dr. Scheuchzer and on *Homo diluvii testis.*" In *Toward a History of Geology*, ed. C. J. Schneer, 193–213. Cambridge and London: MIT Press, 1969.

Kaminski, M. A., et al. "Late Glacial to Holocene Benthic Foraminifera in the Marmara Sea: Implications for Black Sea-Mediterranean Sea Connections Following Last Deglaciation." *Marine Geology* 190 (2002): 165–202.

Karlstrom, K. E., et al. "Model for Tectonically Driven Incision of the Younger Than 6 Ma Grand Canyon." *Geology* 36 (2008): 835–38.

Karlstrom, K. E., et al. "^{40}Ar/^{39}AR and Field Studies of Quaternary Basalts in Grand Canyon and Model for Carving Grand Canyon: Quantifying the Interaction of River Incision and Normal Faulting across the Western Edge of the Colorado Plateau." *Geological Society of America Bulletin* 119 (2007): 1283–1312.

Keill, J. *An Examination of Dr. Burnet's Theory of the Earth, Together with some remarks on Mr. Whiston's New Theory of the Earth.* Printed at the Theater, Oxford, 1698.

Kirwan, R. *Geological Essays*. London: Printed by T. Bensley for D. Bremner, 1799.

Klaver, J. M. I. *Geology and Religious Sentiment: The Effect of Geological Discoveries on English Society and Literature Between 1829 and 1859*. Leiden, New York, and Köln: Brill, 1997.

Kulp, J. L. "Deluge Geology." *Journal of the American Scientific Affiliation* 2 (1950): 1–15.

——. "The Carbon 14 Method of Age Determination." *Scientific Monthly* 75 (1952): 259–67.

Lenormant, F. *The Beginnings of History*. New York: Scribner's, 1883.

Levin, D. "Giants in the Earth: Science and the Occult in Cotton Mather's Letters to the Royal Society." *The William and Mary Quarterly* 45, Third Series (1988): 751–70.

Lewis, M. E. *The Flood Myths of Early China*. Albany: State University of New York Press, 2006.

Ludwin, R. S., et al. "Dating the 1700 Cascadia Earthquake: Great Coastal Earthquakes in Native Stories." *Seismological Research Letters* 76 (2005): 140–48.

Luther, M. *Luther's Works, Volume 2, Lectures on Genesis,* chapters 6–14. Ed. J. Pilikan. St. Louis, MO: Concordia Publishing House, 1960.

Lyell, C. *Principles of Geology, Being an Attempt to Explain the Former Changes of the Earth's Surface by Reference to Causes Now in Operation*, vol. 3. London: John Murray, 1833.

Lyell, K. *Life, Letters and Journals of Sir Charles Lyell*. London: John Murray, 1881.

Mallowan, M. E. L. "Noah's Flood Reconsidered." *Iraq* 26 (1964): 62–82.

Marriott, A. L. "Beowulf in South Dakota." *New Yorker*, August 2, 1952, 46–51.

Mather, K. F., and S. L. Mason. *A Source Book in Geology, 1400–1900*. Cambridge, MA: Harvard University Press, 1970.

Mayor, A. *The First Fossil Hunters: Paleontology in Greek and Roman Times*, Princeton, NJ: Princeton University Press, 2000.

——. *Fossil Legends of the First Americans*. Princeton, NJ: Princeton University Press, 2005.

McAdoo, B. G., L. Dengler, G. Prasetya, and V. Titov. "Smong: How an Oral History Saved Thousands on Indonesia's Simeulue Island During the December 2004 and March 2005 Tsunamis." *Earthquake Spectra* 22 (2006): S661–69.

McCalla, A. *The Creationist Debate: The Encounter Between the Bible and the Historical Mind.* London and New York: T & T Clark International, 2006.

McCoy, F. W., and G. Heiken. "Tsunami Generated by the Late Bronze Age Eruption of Thera (Santorini), Greece." *Pure and Applied Geophysics* 157 (2000): 1227–56.

Miller, A. M. "The New Catastrophism and Its Defender." *Science* 55 (1922): 701–3.

Mitchell, S. G., and D. R. Montgomery. "Polygenetic Topography of the Washington Cascade Range." *American Journal of Science* 306 (2006): 736–68.

Montgomery, D. R., et al. "Evidence for Holocene Megafloods Down the Tsangpo River Gorge, Southeastern Tibet." *Quaternary Research* 62 (2004): 201–7.

Moore, J. R. "Charles Lyell and the Noachian Deluge." *Evangelical Quarterly* 45 (1973): 141–60.

Nicolson, M. H. *Mountain Gloom and Mountain Glory: The Development of the Aesthetics of the Infinite.* Seattle: University of Washington Press, 1997 (1959).

Numbers, R. L. "Creationism in 20th-Century America." *Science* 218 (1982): 538–44.

——. *The Creationists.* New York: Alfred A. Knopf, 1992.

Nunn, P. D. "On the Convergence of Myth and Reality: Examples from the Pacific Islands." *The Geographical Journal* 167 (2004): 125–38.

Olson, L., and H. L. Eddy. "Leonardo da Vinci: The First Soil Conservation Geologist." *Agricultural History* 17 (1943): 129–34.

Oreskes, N. *The Rejection of Continental Drift: Theory and Method in American Earth Science.* New York and Oxford: Oxford University Press, 1999.

Oreskes, N., ed. *Plate Tectonics: An Insider's History of the Modern Theory of the Earth.* Boulder and Oxford: Westview Press, 2001.

Origen. *On First Principles: Being Koetschau's Text of the De Principiis Translated into English, Together with an Introduction and Notes.* Trans. G. W. Butterworth. New York: Harper & Row, 1966.

Paine, T. *The Theological Works of Thomas Paine*. London: R. Carlile, 1824.

Parkinson, W. "Questioning 'Flood Geology.' " *Reports of the National Center for Science Education* 24, no. 1 (2004): 24–27.

Pennock, R. T. *Tower of Babel: The Evidence Against the New Creationism*. Cambridge, MA: MIT Press, 1999.

Pimm, S. L., G. J. Russell, J. L.Gittleman, and T. M. Brooks. "The Future of Biodiversity." *Science* 269 (1995) 347–50.

Plato. *Timaeus and Critias*, London: Penguin Books, 1977.

Playfair, J. *Illustrations of the Huttonian Theory of the Earth*. London and Edinburgh: Cadell and Davies; William Creech, 1802.

———. "Biographical Account of the Late Dr. James Hutton, F.R.S. Edin." *Transactions of the Royal Society of Edinburgh* 5, part 3 (1805): 39–99.

Pleins, J. D. *When the Great Abyss Opened: Classic and Contemporary Readings of Noah's Flood*. Oxford and New York: Oxford University Press, 2003.

Price, G. M. *The New Geology: A Textbook for Colleges, Normal Schools, and Training Schools; and For the General Reader*. Mountain View, CA: Pacific Press Publishing Association, 1923.

Pognante, U., and P. Benna. "Metamorphic Zonation, Migmatization and Leucogranites along the Everest Transect of Eastern Nepal and Tibet: Record of an Exhumation History." In *Himalayan Tectonics*, ed. P. J. Treloar and M. P. Searle. Geological Society Special Publication 74, 323–40. London: The Geological Society, 1993.

Ramm, B. *The Christian View of Science and Scripture*. Grand Rapids, MI: Wm. B. Eerdmans Publishing Company, 1956.

Rappaport, R. "Geology and Orthodoxy: The Case of Noah's Flood in Eighteenth-Century Thought." *British Journal for the History of Science* 11 (1978): 1–18.

Ranney, W. *Carving Grand Canyon: Evidence, Theories, and Mystery*. Grand Canyon, CO: Grand Canyon Association, 2005.

Raup, D. M., and J. J. Sepkoski. "Mass Extinctions in the Marine Fossil Record." *Science* 215 (1982): 1501–3.

Repcheck, J. *The Man Who Found Time*. Cambridge, MA: Perseus, 2003.

Rosen, E. "Was Copernicus' Revolutions approved by the Pope?" *Journal of the History of Ideas* 36 (1975): 531–42.

Ross, D. A., E. T. Degens, and J. MacIlvaine. "Black Sea: Recent Sedimentary History." *Science* 170 (1970): 163–65.

Rudwick, M. J. S. "Charles Lyell Speaks in the Lecture Theatre." *British Journal of the History of Science* 9 (1976): 147–55.

———. *Bursting the Limits of Time: The Reconstruction of Geohistory in the Age of Revolution*. Chicago and London: University of Chicago Press, 2005.

Ryan, W., and W. Pitman. *Noah's Flood: The New Scientific Discoveries About the Event That Changed History*. New York: Simon & Schuster, 1998.

Ryan, W. B. F., C. O. Major, G. Lericolais, and S. L. Goldstein. "Catastrophic Flooding of the Black Sea." *Annual Review of Earth and Planetary Science* 31 (2003): 525–54.

Ryle, H. E. *The Early Narratives of Genesis: A Brief Introduction to the Study of Genesis 1–11*. London and New York: Macmillan and Co., 1892.

Sakai, H., et al. "Geology of the Summit Limestone of Mount Qomolangma (Everest) and Cooling History of the Yellow Band under the Qomolangma Detachment." *The Island Arc* 14 (2005): 297–310.

Schneiderman, J. S., and W. D. Allmon. *For the Rock Record: Geologists on Intelligent Design*. Berkeley: University of California Press, 2009.

Schofield, C. I. *Reference Bible*. New York: Oxford University Press, 1917.

Schuchert, C. "The New Geology: A Text-Book for Colleges, Normal Schools and Training Schools; and for the General Reader by George McCready Price." *Science* 59 (1924): 486–87.

Sedgwick, A. "On Diluvial Formations." *Annals of Philosophy* 10 (1825): 18–37.

———. "Address to the Geological Society, delivered on the evening of the 18th of February 1831." *Proceedings of the Geological Society of London* 1 (1834) 281–316.

Smith, G. *The Chaldean Account of Genesis, Containing the Description of the Creation, the Fall of Man, the Deluge, the Tower of Babel, the Times of the Patriarchs, and Nimrod; Babylonian Fables, and Legends of the Gods; From the Cuneiform Inscriptions*. New York: Scribner, Armstrong & Co., 1876.

Smith, G. A. "Missoula Flood Dynamics and Magnitudes Inferred from Sedimentology of Slack-Water Deposits on the Columbia Plateau, Washington." *Geological Society of America Bulletin* 105 (1993): 77–100.

Stuiver, M., B. Kromer, B. Becker, and C. W. Ferguson. "Radiocarbon Age Calibration Back to 13,300 Years BP and the ^{14}C Age Matching of the German Oak and US Bristlecone Pine Chronologies." *Radiocarbon* 28 (1986): 969–79.

Thomson, K. *The Legacy of the Mastodon: The Golden Age of Fossils in America.* New Haven, CT: Yale University Press, 2008.

Tinkler, K. J. *A Short History of Geomorphology.* Totowa, NJ: Barnes & Noble Books, 1985.

Tolmachoff, I. P. "The Carcasses of the Mammoth and Rhinoceros Found in the Frozen Ground of Siberia." *Transactions of the American Philosophical Society* 23 (1929): 11–74.

Turney, C. S. M., and H. Brown. "Catastrophic Early Holocene Sea Level Rise, Human Migration and the Neolithic Transition in Europe." *Quaternary Science Reviews* 26 (2007): 2036–41.

Vail, T. *Grand Canyon: A Different View.* Green Forest, AR: Master Books, 2003.

Van de Fliert, J. R. "Fundamentalism and the Fundamentals of Geology." *Journal of the American Scientific Affiliation* 21 (1969): 69–81.

Virgili, C. "Charles Lyell and Scientific Thinking in Geology." *Comptes Rendus Geoscience* 339 (2007): 572–84.

Vitaliano, D. B. "Geomythology." *Journal of the Folklore Institute* 5 (1968): 5–30.

———. *Legends of the Earth: Their Geological Origins.* Bloomington: Indiana University Press, 1973.

Waitt, R. B. "Case for Periodic, Colossal Jökulhlaups from Pleistocene Glacial Lake Missoula." *Geological Society of America Bulletin* 96 (1985): 1271–86.

Weber, C. G. "The Fatal Flaws of Flood Geology." *Creation/Evolution* 1 (1980): 24–37.

Wernicke, B. "The California River and Its Role in Carving Grand Canyon." *Geological Society of America Bulletin* 123 (2011): 1288–1316.

Whitcomb, J. C., and H. M. Morris. *The Genesis Flood: The Biblical Record and Its Scientific Implications.* Philadelphia: The Presbyterian & Reformed Publishing Company, 1961.

White, A. D. *A History of the Warfare of Science with Theology in Christendom.* New York: D. Appleton and Company, 1910.

Wilson, L. G. *Charles Lyell, The Years to 1841: The Revolution in Geology.* New Haven, CT: Yale University Press, 1972.

Woodward, J. *An Essay Towards a Natural History of the Earth and Terrestrial Bodyes, Especialy Minerals: As Also of the Sea, Rivers, and Springs. With an Account of the Universal Deluge: And of the Effects that it Had upon the Earth.* Printed for A. Bettesworth and W. Taylor in Pater-noster Row, R. Gosling at the Middle-Temple-Gate in Fleet-Street, and J. Clarke under the Royal-Exchange in Cornhill, 1723.

Woolley, L. "Stories of the Creation and the Flood." *Palestine Exploration Quarterly* 88 (1956): 14–21.

Yang, S.-H. "Radiocarbon Dating and American Evangelical Christians." *Perspectives on Science and Faith* 45 (1993): 229–40.

Yanko-Hombach, V., A. S. Gilbert, N. Panin, and P. M. Dulokhanov, eds. *The Black Sea Flood Question: Changes in Coastline, Climate and Human Settlement.* Berlin: Springer, 2007.

Young, D. A. *Creation and the Flood: An Alternative to Flood Geology and Theistic Evolution.* Grand Rapids, MI: Baker Book House, 1977.

———. "Scripture in the Hands of Geologists (Part One)." *Westminster Theological Journal* 49 (1987): 1–34.

———. *The Biblical Flood: A Case Study of the Church's Response to Extrabiblical Evidence.* Grand Rapids, MI: Wm. B. Eerdmans Publishing Company, 1995.

Young, D. A., and R. F. Stearley. *The Bible, Rocks and Time: Geological Evidence for the Age of the Earth.* Downers Grove, IL: InterVarsity Press, 2008.

Zimmern, H. *The Babylonian and The Hebrew Genesis.* Trans. H. Hutchison. London: David Nutt, 1901.

Acknowledgments

ONCE AGAIN, ANNE PATIENTLY endured stacks of books piled up on our dining room table for months on end. Her insight, suggestions, and ideas greatly improved more chapter drafts than one should really ask a spouse to read. My editor at W. W. Norton, Maria Guarnaschelli, helped me frame and shape the story, and tremendously improved the manuscript. I greatly appreciate her taking on this book and her help in crafting it. Her assistant, Melanie Tortoroli, deserves extra special thanks for helping to focus and streamline the narrative, consistently offering up ways to improve the story. Janet Byrne helped polish the manuscript with insightful copyediting. Several people were instrumental in helping with the artwork. Harvey Greenberg made all the wonderful maps. Véronique Robigou helped conceptualize and sketched exceptional draft figures on very short notice. Alan Witschonke also produced beautiful illustrations. I am indebted to Susan Rasmussen and Jessica Cromheecke for help running down source materials, Rachel Walcott for organizing a trip to Siccar Point, Mike Summerfield for sharing his personal library, Lewis Owen for enduring a trip to the Creation Museum, and Charlotte Schreiber,

Blake Edgar, Ronald Numbers, Roger Wynne, and Art McCalla for reading draft manuscripts and offering suggestions for improvements along the way. Ray Troll's song "Rocks Don't Lie" was inspirational. And I am deeply appreciative of Oliver Korup and the Swiss Federal Research Institute for hosting an extended visit to work on the manuscript. Finally, in addition to finding the book a great home, my agent, Elizabeth Wales, offered timely advice and enthusiasm when it was sorely needed.

I would also like to thank several colleagues at the University of Washington for their companionship and assistance in the field. Among them, Bernard Hallet graciously allowed me to use his photograph of the Tsangpo moraine dams, and Amanda Henck Schmidt translated key conversations described in chapter 1. Alan Gillespie, Allison Anders, and Noah Finnegan also played key roles in uncovering the Tibetan lake story. The opportunity to work with wonderful people in an inspirational landscape is a tremendous side benefit of geological fieldwork.

Naturally, I am particularly indebted to scholars whose works I relied on, especially Norman Cohn (*Noah's Flood: The Genesis Story in Western Thought*), Arthur McCalla (*The Creationist Debate*), Ronald Numbers (*The Creationists*), Martin Rudwick (*Bursting the Limits of Time*), Davis Young (*The Biblical Flood: A Case Study of the Church's Response to Extrabiblical Evidence*), and Dorothy Vitaliano, whose classic *Legends of the Earth: Their Geologic Origins* remains both inspirational and informative. In condensing so much into this book, I have had to telescope the evolution of geological and theological thought into the lives of selected major players. As influential as these key protagonists were, in many ways progress was the fruit of incremental discoveries that contributed to setting the intellectual context for the issues and views of their day. To the reader offended at my neglect to cover his or her favorite historical character(s), I can only plead that a sincere desire to prevent this book from mushrooming beyond appeal to a general readership made such oversights inevitable. And instead of naming a few of the key players in developing the theory of plate tectonics, I chose to respect the extended nature of the network of many individuals who transformed the way we see Earth's dynamic surface.

As this book is intended for a general audience, I abandoned traditional academic footnotes and instead document the source materials I drew upon at the end. Naturally, I encourage the enthused, skeptical, or outraged reader to consult them for additional detail, material, and perspectives. Of course, I alone am responsible for any inadvertent errors and the inevitable sins of omission given the tremendous amount of material—and wildly divergent opinions—on the subjects considered in these pages. And finally, I must confess to taking a few inconsequential liberties in recalling my hike out of the Grand Canyon.

Index

Page numbers in *italics* refer to illustrations.